Other titles available in the *Original* series are:

Original AC Ace & Cobra
by Rinsey Mills
Original Aston Martin DB4/5/6
by Robert Edwards
Original Austin Seven
by Rinsey Mills
Original Austin-Healey (100 & 3000)
by Anders Ditlev Clausager
Original Citroën DS
by John Reynolds with Jan de Lange
Original Corvette 1953-1962
by Tom Falconer
Original Ferrari V8
by Keith Bluemel
Original Honda CB750
by John Wyatt
Original Jaguar E-Type
by Philip Porter
Original Jaguar Mark I/II
by Nigel Thorley
Original Jaguar XK
by Philip Porter
Original Land-Rover Series I
by James Taylor
Original Mercedes SL
by Laurence Meredith
Original MG T Series
by Anders Ditlev Clausager
Original MGA
by Anders Ditlev Clausager
Original MGB
by Anders Ditlev Clausager
Original Mini Cooper and Cooper S
by John Parnell
Original Morgan
by John Worrall and Liz Turner
Original Morris Minor
by Ray Newell
Original Porsche 356
by Laurence Meredith
Original Porsche 911
by Peter Morgan
Original Sprite & Midget
by Terry Horler
Original Triumph TR2/3/3A
by Bill Piggott
Original Triumph TR4/4A/5/6
by Bill Piggott
Original Vincent
by J. P. Bickerstaff
Original VW Beetle
by Laurence Meredith
Original VW Bus
by Laurence Meredith

ORIGINAL PORSCHE
924/944/968

by Peter Morgan

Photography by Simon Clay & Dieter Rebmann
Edited by Mark Hughes

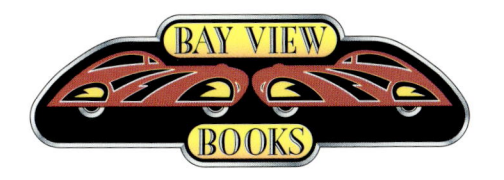

BAY VIEW BOOKS

The photographs on the jacket and preliminary pages
show examples of 968 Club Sport (front cover),
944 Turbo (back cover), 924 Carrera GTS
(half-title page) and 944 (title page).

Published 1998 by Bay View Books Ltd
The Red House, 25-26 Bridgeland Street
Bideford, Devon EX39 2PZ, UK

© Copyright 1998 Peter Morgan

All rights reserved. No part of this publication
may be reproduced or transmitted in any form
or by any means, electronic or mechanical,
including photocopying, recording or in
any information storage or retrieval
system, without the prior written
permission of the publisher.

Designed by Chris Fayers
Sub-edited by Peter Chippindale

ISBN 1 901432 05 X
Printed in Hong Kong
by Paramount Printing Group

Contents

Introduction	6
The 924 (1976-85)	**12**
924 Martini	27
924 Sebring '79	28
924 Le Mans	29
Data Section	31
The 924 Turbo (1979-83)	**34**
924 Carrera GT	43
924 Carrera GTS	45
Data Section	48
The 924S (1985-88)	**49**
924S Le Mans	55
Data Section	57
The 944 (1982-89)	**58**
944 Celebration	72
Data Section	73
The 944 Turbo (1985-91)	**75**
944 Turbo S	82
944 Turbo Cabriolet	84
Data Section	87
The 944S & 944S2 (1986-91)	**88**
944S2 Cabriolet	97
Data Section	99
The 968 (1991-95)	**100**
968 Club Sport	111
968 Turbo S	114
968 Sport	115
Data Section	116
Buying & Driving	**118**
Buying a 924	119
Buying a 944	122
Buying a 968	125
Performance at a glance	127
Engine performance summary	128
Evolution of wheels and tyres	128

Introduction

The 924, 944 and 968 water-cooled four-cylinder models can claim many credits, including contributing one-third of Porsche's all-time production total. Yet there are many who did not mourn their passing in 1995. Nearly 20 years of production had resulted in a love/hate relationship, with the dealers ecstatic about them in the boom years and despairing in the bad ones when they were desperately trying to salvage the heavily tarnished company image.

Customers were the same. Who would not jump at the chance to buy a Porsche, albeit a cheaper version? Those who did generally loved their cars as they basked in the reflected glory of such a famous name. But that fame, another group of loyal customers insisted, had been hijacked. For a 911 buyer, the loss of exclusivity which came from seeing a Porsche on almost every street was often difficult to bear.

The critics had been vocal from the start of the 924 programme, voicing continued doubts about the car's significant VW-Audi parentage. But despite this it still had a good start, especially in Porsche's most important export market of the US. Even though the 924 needed immediate improvement, for a few glorious years production figures went through the ceiling. But the boom did not last long and by 1980 the 924 was on the defensive, under siege from several competitors, notably Mazda, who were producing imitations at

An early evolutionary step: the 924 Martini of 1977 was Porsche's first four-cylinder limited edition and celebrated victory in the 1976 World Championship of Makes.

Introduction

Styling evolution seen in side view (starting from top left): 924, 924 Turbo, 944, 944 Turbo, 944S2 Cabriolet and 968.

much lower prices. Porsche needed to take action.

The 924 Turbo had already begun to address the performance shortfall of the regular car, but the real solution came with the 944. This car finally buried the controversial parentage and, by the mid-'80s, production was booming again as Porsches became *the* fashion items for yuppies – the 'young upwardly mobiles'. In general, all these new buyers wanted was a status symbol which was also a hot sports car – they could not have cared less about Porsche's traditions or where the parts came from. Intoxicated by the enormous demand generated by the 944, Porsche responded by turning out as many of them as possible, with no-one in the company apparently stopping to draw breath, develop new models or even ask what the future held.

Then, after the stock market crashed in 1987, the yuppies suddenly either ran out of money or moved their aspirations on, causing sales to nose-dive. Dealers, finding themselves over-stocked, resorted to price-cutting – something that now became the norm despite never having been heard of before in Porsche circles.

Sadly, the company had made little or no provision for such a rapid down-turn. Although the 911 faltered as well, its more loyal and enthusiastic customers did not desert it as badly as the 944, on whose cash stream the company had become heavily dependent. Worse still, the Porsche name came to signify the worst get-rich-quick excesses of the '80s, giving the company's image a battering. By the early '90s, Porsche was in real trouble, with members of the media having

7

Original Porsche 924/944/968

a field day picking, like vultures, at this tarnished icon of the greed years. Everyone, it almost seemed at one stage, was lining up to kick Porsche while it was down.

Even if new product development had stood still in the '80s, however, some of the profits made out of the 944 had at least been spent wisely. A programme to significantly upgrade production facilities had provided manufacturing flexibility, so that when markets did turn down, the company was able to tune production to orders, most notably by bringing four-cylinder production wholly into the Zuffenhausen plant from 1992.

Yet the early '90s were still years of agony for Porsche. There were many lay-offs, while scores of market analysts gleefully predicted that either

Styling evolution seen from the front (starting from top left): 924, 924 Turbo, 944, 944 Turbo, 968 Cabriolet and 968 Club Sport.

Introduction

Styling evolution seen from the rear (starting from top left): 924, 924S, 944, 944 Turbo, 944S2 Cabriolet and 968 Club Sport.

Daimler-Benz or VW-Audi would buy up the jewel in the crown of the German motor industry. But if one man can take credit for the survival of Porsche as an independent manufacturer to this day, it has to be Wendelin Weideking. Having worked at Porsche in the '80s and then left to join another engineering company, he returned in 1991 to take charge of production – and ruthlessly improved production efficiency. When he became Chief Executive Officer in 1994, he transferred his reforming zeal to the whole organisation, determined to put the company back on top.

Hope sprang alongside the introduction of the 80% new 968 in August 1991. Yet, despite its almost faultless technical specification, the model could not stop the slide. Sales proved to be so

Original Porsche 924/944/968

disappointing that the 968 lasted only three years before it was discontinued, leaving the evergreen 911 to pull the company through until a new generation of cars was developed.

With the winding up of 968 production in 1995, Porsche's association with four-cylinder sports cars came to an end. For enthusiasts, though, these water-cooled models have left a wonderful legacy, the politics and pain now long forgotten in favour of the excitement that they can still generate.

Like the 911, the 924 was not much of a car at first. But like its famous big brother, it went on to become a very good car indeed, peaking with the 924 Turbo and 924 Carrera GT. The 944 which followed was better still, and with the introduction of the Turbo it even matched the 911 as a product, with little to choose, both in performance and looks. And although the 968, which evolved from the 944, may not have starred in the marketplace, it bristled with technical excellence and had undoubted capability.

I have owned six 924s and 944s, and I have driven all the later 968 models in the course of my journalistic work. Yet for me the 924 still has a special place. Some time ago, my wife Anne and I were up to our ears in debt and generally things financial were not looking too great. That was when we bought a beautiful 1979 924 in Colibri metallic green. It did not cost that much, but it looked great and enabled us to drive down the road with our heads held high. That car proved to be the light at the end of our particular tunnel and, when we could afford it, we progressed to a 924 Turbo, discovering that high-performance Porsche driving could be had in a car some ten years younger than a similarly priced 911.

My most recent water-cooled four-cylinder was a 1986 944 Turbo which I owned while living in the US. It might 'only' have had the 220bhp engine, but this, combined with excellent chassis dynamics, still put the car in a performance class of its own. In the wet at the Road America race track, I was even able to drive round the outside of floundering 911s!

Whatever your pocket, or wherever your interest lies, four-cylinder water-cooled Porsches offer incredible value, as well as an opportunity to drive one of the most famous sports car names in the world. Ignore the back-handed comments from the cynics and you will discover performance, build quality and astonishing practicality.

A late evolutionary step: the UK-only 968 Sport of 1994-95 was an outstanding car which combined the Club Sport's engineering with the practicality of a regular four-seater 968.

Introduction

Acknowledgements & Explanations

As always when putting together a text of this detail, I received considerable help from friends. At Porsche AG in Stuttgart, I must thank Olaf Lang and Klaus Parr in particular. I must also record my gratitude to Porsche Cars North America and James Pillar at Porsche Cars Great Britain. I would like to acknowledge the valuable references for this book provided by other authors – Jürgen Barth and Lothar Boschen, Michael Cotton, Karl Ludvigsen, Jerry Sloniger and others. I must also thank friends in the Chicago region of the Porsche Club of America, in particular Nick Brenkus. Sarah Chater translated many heavyweight factory documents for me.

As the trademark of any *Original* title is excellent photography, I must also compliment Simon Clay and Dieter Rebmann on their superb work. We chose cars for photography very carefully to show correct factory specification. As very few of the earlier four-cylinder Porsches have received full restoration in the accepted sense, we relied on cars that have been unusually well preserved, in many cases still in the hands of the original owner and with low mileage. I would like to thank all the owners in three different countries who helped us in this task – they are listed below.

In the UK
Dennis Chorlton (924 Lux); Tim Ewins (924 Le Mans); Veronica Robinson (924); A J Cars of Dibden Purlieu (924 Turbo); Derek Bell (924 Turbo Carrera GTS); Betty Holt (924S Le Mans); Gordon Taggart (924S Le Mans); George Eynon (944 2.7); Andrew King (944 Turbo Cabriolet); Peter Standbridge (944S); Gerry Cooper (944S2 Cabriolet); H R Owen of St Albans (968 Cabriolet); Alan McCrae (968 Club Sport); Meridien Majestic of Lyndhurst (968 Sport).

In Germany
Frank Thiel (924, 924 Turbo, 924 Turbo Carrera GT and 924S); Willi Gerhards (924); Klaus Nierste (944); Wolfgang Kaltschmid (968).

In the USA
Maureen Hutton (924 Martini); Steve Rashbaum (944 Turbo); Lee Lichtenstein (944 Turbo S); Nick Brenkus (944S2); John Mueller (968 Club Sport).

Among the owners who have co-operated in this book is Derek Bell, five times Le Mans winner, who has owned his 1981 924 Carrera GTS since new. This limited-production car grew out of the Carrera GTR, which raced successfully in prototype form at Le Mans in 1980 – Derek brought one of these works cars home in 13th place in that race.

For locating cars, particular thanks are due to Nick Brenkus, Frank Thiel, Stuart Latham, Graham Barber, Fred Hampton and Ray Northway.

When reference is made to a year in this book, it is always the model year – not the calendar year. This is a common motor industry practice that Porsche followed throughout 924/944/968 production. Model years always start in August of the preceding calendar year: the 1978 model year, for example, ran from August 1977 to July 1978. The changes for a new model year were introduced after the factory shut-down for summer holidays.

There is a certain amount of specification overlap between the model families covered in the seven main chapters of this book. To avoid repetition, cross-referencing is used to direct the reader to appropriate information.

The pleasure to be had from any Porsche is above all in the driving – this is a 924S.

Original Porsche 924/944/968

The 924 (1975-85)

In 1972, Ferry Porsche and his sister, Louise Piëch, decided it was time for the family to withdraw from the day to day operations of the company, although it would retain ownership. As a result a new management team, with complete control of future direction, was established under the leadership of Dr Ernst Fuhrmann, while the name of the holding company, Dr. Ing. h.c. F. Porsche GmbH, changed to Dr. Ing. h.c. F. Porsche AG.

Until this point, Porsche cars had been fundamentally inspired by the design philosophy of Ferry's father, Professor Ferdinand Porsche. Now, however, the new managers felt that a change of direction was needed. They believed that the company's mainstay of the 911 had, at the most, only seven more useful years in the market place – thinking which was shortened still further by the world oil crisis of 1973. To them, like many others, not only were expensive sports cars suddenly looking unfashionable, but the 911 itself was compounding the problem by beginning to look very dated (it had been launched in 1964) and lacking in key technical refinements.

Studies had already been progressing to establish what was required in a new sports car, and a replacement model, the 928, was now approved at the new development centre at Weissach, 15 miles west of the Zuffenhausen factory. In a major development, the 928 broke the traditional air-cooled lineage, in line with conclusions that the new car should incorporate several important features – a water-cooled front-mounted engine, a transaxle chassis layout (with the transmission located over the rear axle to provide balanced weight distribution) and the option of a conventional fully automatic transmission.

These new philosophies were then discussed with VW-Audi. The two companies had already achieved excellent results with their joint venture, the VW-Porsche 914, and intended to build on this by taking even more advantage of

Evolution Outline

Nov 1975	Production starts of 125bhp model for Europe.
Apr 1976	First US deliveries, 95hp SAE (as 1977 models).
Feb 1977	'1977.5' models in US get 110hp SAE.
Mar 1977	Deliveries start for right-hand drive versions.
Aug 1977	Improved rear suspension, significantly enhancing ride quality.
Aug 1979	All models receive breakerless ignition, five-speed 016 gearbox.
Aug 1980	Seven-year anti-perforation warranty, problems with hot starting rectified.
Jul 1985	Production ends.

German-registered 924 from 1977 shows simple lines of the tail, with no spoiler on early cars – a benefit when reversing. Pre-1980 side windows had chromed trim, while the round exhaust tailpipe was an easy way to identify a pre-1978 model.

924

A 1979 model with the Lux specification offered in the UK – this included various extras, such as electric door mirrors, as standard.

VW's high-volume methods. A change of leadership at VW in 1971 had also injected fresh enthusiasm for a new VW sports car.

The 914's mid-engined design had dictated a strict two-seater layout, with the car's dynamics also meaning it needed a skilful driver to handle it on fast bends. The new car, it was decided, would need to be more predictable, while remaining agile. VW, too, also wanted a practical 2+2 which met expectations of a water-cooled engine with a conventional front-engined layout. For Porsche, the transaxle solution, with the gearbox mounted on the rear axle line, was the obvious and technically interesting route.

In early 1972, the studies with Porsche were formalised into development contract EA425. Design and styling development then proceeded rapidly, entirely performed by Porsche at Weissach but fully backed by VW. However, other factors were now intervening which were not entirely what Porsche wanted. The joint VW-Porsche marketing and sales organisation, which had sold the 914, had disappeared and VW now insisted the new sports car would be sold only under its badge. Porsche, impressed by the magnitude of the 914's sales of more than 118,000 in six years, did not want to hear this at all. The new managers at Porsche had realised that this market sector remained unexploited and could never be approached by the upmarket 911, or even its eventual replacement. Their dealers wanted a new, volume-selling model.

Yet, by the end of 1974, when eight prototypes were up and running, things had changed once again. The world oil crisis had drastically altered the shape of the automobile market and the EA425, with by now over $50 million spent on design and production start-up, had dropped out of VW's vision of the future. Porsche, however, remained bullish about its prospects and took advantage of another change of leadership at VW to make the decision to buy out the project. This was an extremely bold move, especially as it was taken when sports car sales were flat.

The deal was concluded in January 1975 for more than $40 million and included design, tooling, marketing and sales. However, as Porsche did not have the production facilities which would be needed, it signed a contract with VW-Audi for them to build the new car, but under Porsche quality control, in the old NSU factory at Neckarsulm, north of Stuttgart.

Original Porsche 924/944/968

Development proceeded rapidly during the year, with the EA425 being given the visual identity, performance and levels of equipment which customers expected of a Porsche. During the summer around 100 pre-production cars came off the Neckarsulm line, with full production starting in November. The designation of type 924 – a number rather than a name – was chosen entirely for marketing reasons and had no internal significance to the company.

After what must have seemed a slow first year, sales of the 924 took off. At first the stumbling block had been the performance of the US models, which only claimed 95hp SAE and therefore created a credibility problem for a car carrying the Porsche badge. But a mid-year revision in 1977, which had started with the first right-hand drive cars being delivered to British customers, sorted that out to a large extent, after which US sales beat all expectations. Whereas only 5145

The pop-up headlamp design – and indeed the whole profile of the nose – was to become much copied by Japanese manufacturers. Guards Red suited the 924 perfectly, and came to be regarded as almost the stereotype colour for the four-cylinder models.

From the 1982 model year the Turbo's rear spoiler was fitted to the 924, greatly improving the appearance from the rear.

924

This 1983 model has factory-fitted side decals and sunroof. Owners of right-hand drive 924s always had to cope with windscreen wipers that parked on the 'wrong' side.

The opening all-glass tailgate, an advanced feature for 1975, dramatically improved the practicality of the car. A disadvantage, however, was that the large glass area made the interior very hot on sunny days.

924s had been produced in 1976, total production more than quadrupled to 23,889 in 1977.

As soon as it was launched, the 924 was subjected to a typically Porsche period of intense development aimed at losing the unpopular association with VW. By 1980 and, ironically, when the new Audi five-speed gearbox was fitted as standard to all models, the 924 could boast a level of refinement way beyond the first cars. A year later, in 1981, the 100,000th production model came off the Neckarsulm line. Yet, almost exactly at this point, fashion was moving on and sales in the US had begun to decline.

The new 944, therefore, arrived not a moment too soon by replacing the 924 in the US in the spring of 1982. In other markets the 924 was to survive a further three years, benefiting from many of the 944's improvements, until, when production finally ceased in 1985, nearly 120,000 had been manufactured in just under ten years.

By any standards, for an independent sports car manufacturer the 924 had been a great success.

Bodyshell

Porsche stylist Harm Lagaay is credited with evolving the original lines of the 924, which was described as 'a modified wedge inspired by the earlier rounded features of the 911 and 356'. However, the new car did not have the same aggression in its shape as the 911, a change which was considered necessary to make it more acceptable to a wider market. During the transition from VW to Porsche, the lines became less rounded and wind tunnel testing enhanced the original 'wedge' concept by producing a windscreen raked back at 60°, along with a small lip spoiler at the front.

One of the most distinctive features of the car was the clean line of the plunging nose, achieved by using pop-up headlights, scrapping the front grille and tilting the engine onto its side. Meanwhile, although the rear side windows were fixed, the opening rear glass was a unique innovation as it was so large and also frameless. Owners of 924s in hot climates would later criticise the 'greenhouse effect' that this produced, but stylistically it was undoubtedly one of the car's successes.

The body was a unitary construction in steel, with the main components carefully laid out to utilise the full advantages of the transaxle gearbox by achieving a near-ideal front/rear weight distribution of 53%/47%. It was this inherent good balance which gave the 924 and all its subsequent derivatives such excellent chassis dynamics.

The coupé bodyshell was also very stiff, thanks to thick A and B pillars, deep sill boxes and a high central tunnel. Customers could also specify a large, glass-fibre lift-out roof panel, which actually gave greater bodyshell stiffness because of the strengthening structure around the opening.

Original Porsche 924/944/968

To make replacement easier, front wings were bolt-on. The rear-mounted fuel tank began with a capacity of 62 litres (13.6 Imperial gallons, 16.4 US gallons), which was then increased to 66 litres (14.5 Imperial gallons, 17.4 US gallons) for the 1980 model year. Approximately 6 litres of this was for the reserve.

There were comparatively few bodyshell changes throughout the ten-year life of the car. For the 1979 model year the internal rear wheel arches were opened up slightly to allow better clearance when using snow chains. For 1982 the roof was strengthened to enable loads of up to 75kg (165lb) to be carried using the 'Porsche Carrier System'. For 1983 gas struts were fitted to keep the bonnet open.

The main attention instead was focused on improving the quality of build, paint finish and trim. Early production vehicles had used hot-dipped zinc coating for both the bodyshell's load-bearing and underbody areas, but for the 1978 model year the entire shell was zinc-coated. From the beginning Porsche offered a six-year anti-corrosion guarantee against rust perforation on the underbody areas, but from the start of the 1981 model year this was increased to seven years and now covered the entire bodyshell.

Body Trim & Fittings

Because the 924 had started by being developed for VW, the original specification, and particularly the trim, was extremely basic. Wherever possible, VW parts had been built in, most obviously in the door handles (and their mechanisms) and fuel filler cap taken from the Golf/Rabbit. To compensate, many markets equipped the basic factory-specification models with 'standard extras', producing such variations as the British Lux and the US Touring Packages (see options list, page 32). During early production in 1975, the trim specification was upgraded to include tinted glass and chromed side window surrounds, with side rubbing strips added from August 1976.

For most markets the bumpers were formed principally from glass-fibre and mounted onto deformable structures attached to the body. They were clearly derived from the 911's successfully integrated units and featured, at the front, recessed driving lights as well as turn indicators. For the US market, the bumpers were manufactured from aluminium – the front a casting and the rear a large extrusion – and then mounted on telescopic energy absorbers.

Wind tunnel development had resulted in rain guttering to keep the glass clear around the top and sides of the bonded windscreen (another first for Porsche), with a recessed channel also running ahead of the opening rear glass. The clip-on guttering at the sides also covered the welded body seams. The door mirrors were low-cost VW parts, with the one for the passenger door only available as an option. However, for 1978 the electrically operated and heated mirrors from the 911 became optional, and then standard for 1980.

One detail which was never rectified on the 924 was the windscreen wipers sweeping the 'wrong' way for right-hand drive cars, resulting in a significant unswept area just to the right of the driver's view ahead.

The first cars used the heating and ventilation system from the VW Golf/Rabbit. Fresh air was drawn in from a large plenum chamber just ahead

Door handles came from the VW Golf/Rabbit, initially chromed but for most of the 924 period with a black anodised satin finish.

Early cars left the factory with simple but effective non-powered external mirrors (far left). Electrically operated and heated mirrors (left) became optional for 1978 and standard for 1980.

The early 'cog wheel' fuel filler cap (far left), straight from the VW Golf/Rabbit, was replaced by a much more elegant concealed filler (left) for 1980.

924

Different front bumper and lighting arrangements for European (right) and US (far right) cars. European glass-fibre bumper carried neatly recessed driving lights, which were displaced by rubber inserts on the larger, impact-absorbing, aluminium bumper for US.

Tail comparison shows similar bumper contrast between European (right) and US (far right) cars.

Rear wiper was located on the left-hand side irrespective of market. With bonded-in rear light lenses, bulbs were replaced from inside the luggage area. The flexible rear spoiler, optional for 1982 and standard for 1983, improved the looks of the 924 but reduced rearward visibility, especially when parking. The drain holes should be kept clear of debris.

This duct in the door shut-face (far right) carried air from the rear of the car through the doors and out to the wheel arch.

Tail badge graphics were changed for 1978. Later cars announced the manufacturer's name far less discreetly – with the word 'PORSCHE' in big outline letters across the lower part of the tail.

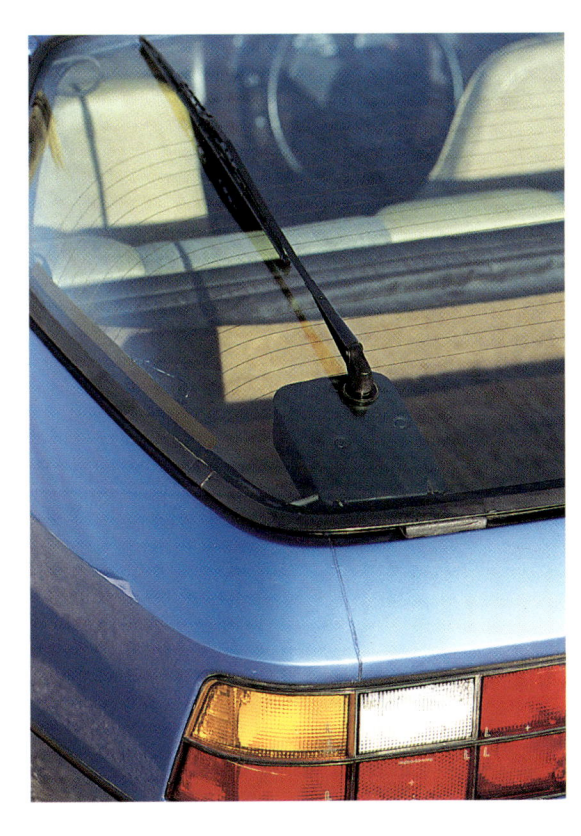

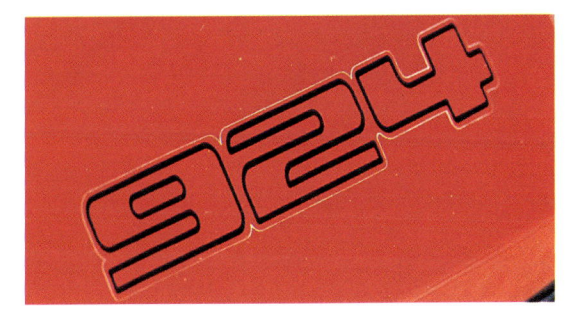

Original Porsche 924/944/968

of the windscreen and extracted through vents below the rear glass. The exhausted air was then drawn forward along each side of the body, through the doors, and vented at the door fronts. From 1979, ventilation was improved by adding separate fresh air blowers, while for 1982 the system was significantly upgraded by introducing greater capacity, a more powerful fan and improved heating. The new design also integrated the optional air conditioning more effectively into the ventilation system. Yet it still failed to be a complete solution, as heat output was difficult to control and the central vent pumped out warmth, even when not required.

For 1980 the 'cog-wheel' VW fuel filler cap was replaced by a new lockable filler hidden behind a body-coloured plastic flap. The progressive elimination of VW parts continued with the adoption of 928 door locks, while from 1981, along with all other Porsches, small rectangular side indicator repeaters were fitted on the front wings. In 1982 fuel tank ventilation was improved.

In 1980 two-tone paint schemes were introduced, with the side window trim being changed from chrome to a black anodised finish. Also long overdue in improving the car's looks was the Turbo's rear spoiler, introduced for 1982 as an option, then becoming standard for 1983.

The interior was neat and practical, if not lavishly furnished, and the rear seats offered satisfactory accommodation for children, or adults on short journeys. These cars show early styles of cloth seat inlay in Tartan Dress and Herringbone.

Interior Trim

In order to upgrade the specification from VW to Porsche standards, interior detail was heavily revised during 1975. The original VW steering wheel was replaced by a somewhat dull 380mm (15in) item, with the rim offset to give more leg room and an improved view of the instruments for tall drivers. The chunky three-spoke design used on the 911 was available as an option but became standard for the 1982 model year. Yet, if one single criticism could still be levelled at the 924's interior, it was the poor leg room provided below the steering wheel.

924

Two more interiors show cloth upholstery choices for the later 924 years. Pinstripe (right) is again confined to seat inlays, but popular Berber (far right) covers the entire seat facings. The most striking offering, Pasha fabric, is shown on page 33.

Porsche crest was added to glove compartment catch for 1982.

Door trim architecture remained basically the same throughout the 924's life, but details changed and equipment improved: 1977 car (right) has stark vinyl with manual windows; 1979 car (far right) has a cloth panel and optional electric windows; 1981 car (below right) gains stridently embossed door tops; and 1983 car (below far right) reverts to all-vinyl trim and plain door tops, but with carpet on the pocket!

To begin with little could be done about the bland plastic facia, originally styled for VW, as it was already tooled and ready for production. Instead the material of the door trim was upgraded and a centre console added housing heater controls, engine instruments and the radio. Manual window lifters were retained, with electric ones an option for the 1978 model year. The lower cost loop-weave carpet originally specified by VW was replaced by higher quality pile, along with cloth-trimmed, high-backed seats taken from the 911 and now offered in a choice of

Original Porsche 924/944/968

Earlier cars used a dull two-spoke steering wheel with the rim off-set slightly above centre to improve leg-room and instrument visibility. The sweep of the rev counter needle, all the same, is arranged so that the engine's working range is not hidden by the rim! Close-up shows that the speedometer was marked to 150mph instead of 250kph for appropriate markets, and this car reveals itself as a US model by its Exhaust Gas Recirculation (EGR) warning light.

The three-spoke steering wheel design used on the 911 was initially an option, then standard for 1982. Without rim off-set, this wheel concealed more of the outer instruments, including the water temperature reading, while a chunkier wiper stalk and re-aligned rev counter combined to obscure the important mid-range revs! Instrument graphics were different and new speedometer readings, perversely, went to 160mph (up 10mph) or 240kph (down 10kph).

tartans and herringbone patterns. From January 1978 leather seats became an option and a leather gaiter was fitted to the gear lever.

VW's drive to reduce costs had also resulted in a fairly noisy interior, but again there had been little time before launch to do much about this. Build quality, also, was not what Porsche owners had come to expect and early 924s soon gained a reputation for rattles, along with poor electrics.

In the continuous subsequent effort to improve comfort, appearance and particularly the level of interior noise, materials became more what was traditionally expected of Porsche, not just better quality, but more adventurous and refined. The loud Pasha chequerboards and the tasteful pinstripes introduced for the 1980 model year were an infinitely more attractive choice than the previous tartans and bleak blacks and browns. More sound insulation followed for 1981, along with carpeting on the centre console. The following year came the nice addition of an attractive gold anodised Porsche crest on the glove compartment door – which proud owners could fit to their earlier cars – along with 'Porsche' embossed in the top of each door trim.

In 1982 the 924 moved to share its interior with the 944, ushering in new carpeted door pockets and the extremely popular Berber fabric, with the discreetly lettered 'Porsche' fabrics offered as an option in 1984.

Overall, these were significant improvements to interior finish and noise reduction, making early '80s cars very attractive propositions for the enthusiast on a budget.

The large luggage space could hold any number of strangely shaped objects. The seat backs could fold forward to give more room, or the roller blind cover pulled out to cover the main stowage area. With the carpet lifted, but not the sound insulation, spare wheel is revealed, and the jack can be seen fitted to the rear wall behind it. The jointed plug socket in the basic tool kit was essential to remove the inaccessible spark plugs.

Dashboard & Instruments

The grained plastic dashboard and sculpted side mouldings were the product of the Porsche styling studio, although all instruments and controls originated from VW. The three main instruments, carrying white lettering and luminous orange telltales, were a variation on the 911's famous 'port holes', but much more strongly emphasised. Unusually, each instrument had a conical cover to reduce reflections while, unlike the 911, the speedometer was in the centre, with the tachometer on the right and combined fuel and water temperature on the left, with their warning lights.

The centre console, a late addition in 1975, significantly improved the overall feeling that the car was indeed a grand tourer. This console housed an analogue clock and oil pressure and temperature gauges above the heater controls, which were awkward to reach past the stubby gear lever. Behind the lever was a recessed ash tray, and there was a small shelf for oddments.

From the start of the 1981 model year, the steering column switch gear was improved.

Luggage Compartment

One of the most enduring features of the water-cooled four-cylinder series was the brilliant luggage compartment. The opening rear glass, combined with a rear seat back which folded forwards, gave huge load-carrying capacity. Along with the small rear seats, this luggage space enabled the 924 to boast correctly that it truly was a car for the young family.

The rear carpeting lifted at each side to reveal large storage wells behind each wheel arch. Lifting out the carpet completely exposed the spare wheel, which was contained, along with the jack and the wheelbrace, in a deep recess between the rear light clusters. The rear lamps had to be accessed from inside the luggage compartment, as the large external lenses were bonded into place.

From August 1976, as well as luggage tie-downs, a useful roller blind was fitted to the folding rear seat back, giving extra security by concealing items left in the rear.

Engine

There was a great deal of discussion before the 924's new engine was finalised. In his autobiography, Ferry Porsche hinted that in the early '70s he considered the then-forthcoming Audi five-cylinder engine as a good option. This was the idea of his ever-creative nephew, Ferdinand Piëch, who had brought considerable flair and not a little argument during his nine years as head of Research and Development at Porsche. After the family had withdrawn in 1972, he had moved to a similar position at Audi, while still maintaining close ties with Porsche. This led to the suggestion that two five-cylinder engines could be combined

Original Porsche 924/944/968

Snug fit! The 2-litre engine was canted 40° over, with the spark plugs on the awkward side, on the left.

to make a V10 for the 928. But apprehension at Porsche that it would become too closely tied to VW-Audi meant the idea was never pursued. Instead Porsche designed a new V8 for the 928.

There were reports confirming that an extremely rapid prototype 924 had been built with a mid-mounted 911 engine, yet in the end it was another Audi engine, a 2-litre four-cylinder manufactured at VW's Salzgitter plant, which was chosen for what was to be a mid-market, rather than mid-engined, car. This engine originally dated from an 1871cc (114.2cu in) pushrod unit first built in 1965 and then extensively redesigned by Audi in the early '70s as Project EA831.

The principal improvements were the introduction of a belt-driven overhead camshaft and an increase in bore from 84.0mm (3.30in) to 86.5mm (3.41in), giving a swept volume of 1984cc (121.1cu in). Stroke remained the same at 84.4mm (3.32in). The increased bore also resulted in the cylinders being siamesed together, a new front-mounted water pump ensuring that coolant found its way along both sides of the iron block. Shortly after the start of production the five main crankshaft bearings were increased from 60mm (2.36in) to 64mm (2.52in).

Because the engine would be tilted 40° to fit under the low bonnet line, a special heavily-finned cast aluminium sump pan had to be manufactured. The finning meant that no separate oil cooler was needed, with the oil pumped by a new crescent-type unit driven from the front of the crankshaft. For Porsche's version of this engine the crankshaft was forged rather than cast.

The new camshaft was supported in five plain bearings in an equally new cast aluminium head, and then driven from the front of the crankshaft by a toothed belt. The camshaft operated bucket-type tappets which included tapered screw adjusters and, for the exhaust tappets, a slight centre offset relative to the cam lobe to encourage rotation and so reduce wear. There were two valves per cylinder, diameters being 40mm (1.57in) for the inlet – or 38mm (1.49in) on US models – and 33mm (1.30in) for the exhaust. The valves operated into a flat heron-type combustion chamber by means of double coil springs.

The cast pistons featured dished crowns, with indents to allow the valves to clear should the camshaft drive belt fail. For the 1977 models the connecting rods were given oil drillings to improve piston cooling.

The distributor was mounted directly off the rear end of the camshaft. US models were fitted

A late model engine bay to 125bhp European specification. The view above shows how the distributor is driven from the rear end of the camshaft, and the headlamp lifting crank is also visible at upper right. The other view shows the intake arrangements and, at lower left, the Bosch K-Jetronic fuel metering unit.

with breakerless transistor ignition from the start, while other markets were given traditional coil ignition until this was dropped with the 1980 model year. Despite the 40° inclination of the engine, the spark plugs were mounted on the exhaust side, and consequently were extremely difficult to get at.

For Porsche's installation, the engine used Bosch K-Jetronic fuel injection, with fuel introduced directly above the inlet valve of each cylinder. The metering unit fed incoming air through double throttle chokes into individual ram tubes and then the cylinder head. Improved mid-range response was achieved through the two-stage opening of the throttle chokes, with the second butterfly not opening until the first was open 50°.

Early cars suffered from fuel vaporisation in heavy traffic on hot days – a problem it was said the factory test engineers had never noticed as they always drove the prototypes flat out! For the 1977 year, the fuel pump was moved to the right-hand rear wing (to give better cooling) and a second fuel pressure accumulator was added. The following year also ushered in a magnetic one-way valve, which significantly improved starting when the engine was hot, with the last improvement to finally solve the problem the addition of a second fuel pump, inside the fuel tank, in 1980.

The fuel pump had also always been criticised for being noisy and for the 1981 model year a new Kurzhals EKP pump was used. This had an integrated pressure retention valve to prevent the fuel in the pipes siphoning back into the tank, thus causing air to enter the system. The same year also saw the introduction of steel injection pipelines and improved cold start control. For the US market, the Lambda control was improved and an active carbon filter used on the fuel tank breather.

The water cooling system was conventional by industry standards, although it used an electric cooling fan, which cut in at 92°C and switched off at 87°C. Another sensor kept the fan running if the engine was stopped and the temperature remained high.

Most models had a three-box exhaust system, with California models fitted with catalytic converters. Cars going to other US states also had air injection, while all US models had the much-hated exhaust gas recirculation, in order to comply with the new emissions standards. For the 1978 model year the original small round tail-pipe was replaced by the more attractive oval one.

With 125bhp the 924 had perky, if not brilliant, performance. Yet the first American versions suffered greatly from both their emissions equipment and the lower compression ratio – 8.0:1 instead of 9.3:1 – which had to be used to suit lead-free fuel. As a result they produced only 95.4hp at 5500rpm (measured by the SAE standard), along with dismal maximum torque of 109.2lb ft at 3000rpm.

Unsurprisingly, the improvements began immediately. From February 1977 the US cars had their compression ratio raised to 8.5:1 and were given revised timing and a new camshaft. This increased power significantly, to 110hp SAE at 5750rpm, although the gain in maximum torque was disappointing with only a slight rise to 111.3lb ft at 3500rpm. As this upgrade came near the half-year point, these improved US cars are referred to as '1977.5' models.

Transmission

The four-speed Audi 088 transmission, built at the VW plant in Kassel, was mounted in line with the rear axle, although cost prevented the use of the excellent Porsche-patented synchronising ring mechanism fitted to the 356 and 911, and which, at one stage, had even been used on early VW/Audi gearboxes.

There were special gears with lower first and second gear ratios to improve low-speed acceleration, along with a higher final drive. The 125bhp models had a rear axle ratio of 3.444:1 (9/31), but the lower powered US models had 3.888:1 (9/35) to improve acceleration.

The single cable-operated 215mm (8.46in) diameter dry-plate clutch remained at the rear of the engine, with torque transmitted to the gearbox by a welded steel tube assembly. The tube, 85mm (3.35in) in diameter and with 4mm (0.16in) walls, carried a drive shaft of 20mm (0.79in) diameter with splines at each end. The whole unit was rigidly mounted to the clutch housing at the front and the gearbox bellhousing at the rear. To minimise noise and vibration, very great care was taken to support this drive shaft on four bearings, with ball bearings supported by rubber rings inside the outer tube. From the 1978 model year a vibration damper was added.

The type 0-87 three-speed automatic transmission, complete with torque converter, was a standard VW-Audi assembly, again built in Kassel, but with ratios specific to the 924. The complete unit was located on the rear axle line using a special torque tube, this time with only three support bearings. Customers paid a price in performance, but at least the automatic was what they generally expected, unlike the 911's Sportomatic system, a clutchless semi-automatic which still required the driver to move a conventional lever to change gear. The somewhat pedestrian three-speed automatic lasted the entire life of the car, the only improvement being to the torque converter in 1982 to give a smoother shift and reduce acoustic noise and fuel consumption.

From June 1977, the Porsche/Getrag 016 five-

Gear levers for the two Audi-designed manual gearboxes used as standard in 924s. The four-speed was replaced by the five-speed for 1980. Gaiters are also different, respectively in rubber and leatherette.

The front suspension (below) was a well-proven MacPherson strut arrangement. Note the solid disc and the front anti-roll bar. Rear suspension (bottom) was by the familiar Porsche principle of semi-trailing arms working transverse torsion bars. Note the large bump stop and rear-mounted gearbox location – the secret of the 924's excellent handling.

speed manual gearbox was offered as an option, with factory brochures claiming that the extra gear reduced the 0-100kph (0-62.5mph) time by 0.3sec. Final drive ratio of the 016Z (for most markets) was 4.714:1 (7/33) while the 016Y (for the US) was 5.0:1 (7/35). This 'box was much better for cross-country and high-speed driving, but its dog-leg first gear was frustrating in heavy traffic. For the 1980 model year, therefore, Porsche introduced a new Audi-designed five-speed gearbox, the type 016.8 or 016.9, which had first and second in line. Also built in Kassel, this 'box had its gear cluster behind the differential, whereas the gears in the Getrag were ahead of the axle line. Final drive was 3.89:1 (9/35) on the 016.8 (for most markets) or 4.11:1 (9/37) on the 016.9 (for the US).

Although this new 'box possibly reduced the agility of the car through fast corners, due to the higher polar moment of inertia created by its rear-biased configuration, otherwise it was precisely what the 924 had needed from the beginning, combining smooth, positive action with a traditional H pattern for the first four speeds, with fifth and reverse off to the right.

For 1982 both synchromesh and gearbox ventilation were improved and for the first time a limited slip differential, with a 40% locking factor, was offered as an option. Further revision for 1983 brought synchromesh to reverse, while the specifications of the automatic transmissions for the 924 and 944 were standardised.

Electrical Equipment

The 924 used a 12-volt electrical system with, it sometimes seemed to frustrated owners, a temperamental relay for every function. Manual cars were fitted with 1050 watt (75 amp) alternators and 45Ah batteries, while automatics had 63Ah batteries – which were in addition an option on the manuals. From the beginning the rear glass was heated, although a rear wiper (always on the left-hand side) was an option.

From August 1976 a volt meter replaced the oil temperature gauge in the centre console, while for 1978 electric window lifters and a three-speaker Porsche radio/cassette became options, along with an electrically operated aerial. From 1980, the interior lighting was upgraded and improved so that the luggage compartment was lit when the rear glass was opened. Externally, a rear fog light was fitted from 1981, along with the twin air horns which had previously been confined to the Turbo. Stereo control was also improved by adding a fader for the dashboard speaker.

A new generation of radio was brought in for the 1983 model year with the Blaupunkt Monterey SQR22 for the US market and the Köln SQR22 for elsewhere, along with provision for a four-speaker stereo system. For 1984, the 924 received the 944's electrically-tilted roof mechanism and rear glass release through a switch mounted on the centre console. The last changes, for the 924's final year, were heated windscreen washer jets and a graduated tint to the top of the 'screen itself.

Suspension & Steering

At the front the 924 used conventional MacPherson struts with coaxial coil springs and shock absorbers (taken from the VW Beetle) in conjunction with fabricated lower A arms (from the Golf/Rabbit). Porsche's engineers then built in the Audi concept of negative roll radius to ensure the car stopped in a straight line on surfaces with uneven grip, or if one front brake failed.

Original Porsche 924/944/968

The steering rack was the Golf's ZF, fitted with a higher ratio of 19.2:1 (rather than 17.2:1) to give four turns lock to lock. The rack was mounted ahead of the front wheels while, to give increased crash protection to the driver, two universal joints were built into the angled column, along with a collapsible section below the dashboard which was designed to reduce potential injury.

The rear suspension had the familiar Porsche look of independent transverse torsion bars operated from semi-trailing arms, with either Boge or Fichtel & Sachs shock absorbers. Drive from the gearbox was through universally jointed half shafts. In a major redesign for the 1978 models, the whole rear assembly was rubber-mounted to the body, significantly reducing interior noise, while Koni shock absorbers were now available.

At the start anti-roll bars were options, with diameters of 20mm (0.79in) at the front and 18mm (0.71in) at the rear. Then came a succession of changes. For 1978, the front bar was stiffened to 23mm (0.91in) and the rear bar softened to 14mm (0.55in). For 1981, a 20mm (0.79in) front bar became standard, with the rear torsion bars stiffened to compensate. For 1983, the front bar was stiffened to 21mm (0.83in), making the already well-balanced 924 superb for fast road use.

In addition, for 1982, the front wheel castor was reduced from 2° 45' to 2° 30', making the steering lighter.

Brakes

The braking system was based around a diagonally divided dual circuit, with solid discs at the front and drums at the rear.

The front discs had a single piston and used fully floating calipers from the Audi 100. The discs themselves had a diameter of 256.5mm (10.10in), with the stamped steel caliper containing a single 48mm (1.89in) piston. The rear drums, taken from the VW Beetle, had a diameter of 228.6mm (9.00in) and were 38.1mm (1.50in) wide. The handbrake, located beside the driver's door on both left-hand and right-hand drive models, operated directly onto the drums.

The braking system was boosted by a servo, giving a soft feel to what were already weak brakes by Porsche standards. The servo was originally of 178mm (7.00in) diameter, but for the 1980 model year a larger version of 228.6mm (9.00in), the same size as the 924 Turbo, was introduced, and the master cylinder was also enlarged from 20.64mm (0.81in) to 23.81mm (0.94in).

The handbrake was located between the sill and the driver's seat, at just the right height to catch on clothing!

Wheels & Tyres

From launch the 924 was equipped with 5.5J × 14 ventilated steel wheels with four mounting studs and carrying 165HR14 tyres. From January 1976, Porsche-designed eight-spoke 6J × 14 cast alloy wheels were available as an option, carrying 185/70HR14 tyres. For the 1978 model year these were also available with glossy black centres, and there was the new option of 6J × 15 wheels carrying 205/60 HR15 tyres, although these made the steering extremely heavy.

Porsche-designed eight-spoke cast alloy wheels became available as a factory option two months after launch, but in most countries they were fitted as standard. The normal finish was natural alloy, but glossy black centres were available from 1978 and the Martini limited edition had an all-white finish.

26

The 1977 Martini was the first 924 limited edition, and has become the most sought-after of the early models. All were finished in Grand Prix White (a 911 colour) with eye-catching red, blue and black Martini stripes.

From the start of the 1981 model year the multi-spoke wheels which had been offered on the 924 Turbo were made available in four-stud form. Black powder coated and with bright turned spokes, they were fitted with 205/60HR15 tyres and gave a fresh look to the 924 from the side. Owners, however, moaned that a toothbrush was needed to clean them properly. Their lacquered finish was also prone to chipping, causing them to corrode when coated with brake dust.

From 1980, in most markets, the spare wheel was the collapsible Space Saver type. The UK was an exception, but adopted it later.

924 Martini (1977)

The first attempt at a limited edition 924 was factory order number M426, celebrating Porsche's victory in the 1976 World Championship of Makes. These cars were painted in Grand Prix White, with the cheerful blue, red and white race side stripes of sponsor Martini & Rossi. Each was given a dash plaque recording the championship victories of 1969, 1970, 1971 and 1976.

Specially white-painted 6J × 14 steel wheels were fitted with 185/70HR14 tyres and there were anti-roll bars front and rear. Otherwise the suspension remained standard, as did the engine.

Inside, special bright red carpets were highlighted by matched fabric inlays on the black leatherette seats, which also featured blue piping. The seat headrests had vertical Martini red, blue and black striping.

Martini 924s are very difficult to find now, and the package looks pretty mild compared with later production models. But these models were the first attempt to perk up the 924's image by identifying it with the company's highly successful racing activity.

Of all the variants with the four-speed Audi gearbox, this was the one to have.

Special French Model (1978)

Beginning in September 1978, 100 special 924s, known as M427, were produced for the French market. They were finished in Dolomite Grey metallic paint, with black and silver 'cheat lines' along the sides. The interior was black with leatherette and pinstripe velour seats.

There were two different equipment specifications. F01 had a driver's electric door mirror, rear wiper, tinted windscreen and side glass, and rear fog light. F02 consisted of F01 plus anti-roll bars front and rear, two stereo speakers, a manual aerial and wiring for a radio, light alloy 6J × 14 wheels with black centres and 185/70HR14 tyres, a leather steering wheel and electric windows.

Original Porsche 924/944/968

924 Sebring 79 (1980)

To celebrate Porsche's victory in the World Championship of Makes, along with five other international and 25 national championships, 1400 Sebring '79 924s, designated M429, were offered in the US only. They were built from December 1978 to February 1979 and garishly finished in Guards Red, with bold yellow, red, black and white side stripes, along with 'Sebring 79' script ahead of the doors and an enormous '924' legend on the nose.

The cars came loaded with what had previously been optional extras, including a removable sunroof, three-speaker stereo, electric exterior mirrors on both sides and black-trimmed fog lights. Like the earlier Martini 924, they had front and rear anti-roll bars, but this time the popular 6J × 14 black-centre alloy wheels were specified, fitted with 185/70HR14 tyres. Sadly for enthusiasts, the standard 110hp SAE was not improved on, but there was the choice of the five-speed Getrag manual gearbox or the three-speed automatic transmission.

Inside there were special orthopaedically designed jet-black bucket seats, with red/blue tartan check inlay from the 911 also echoed on the door panels. Carpet was red.

Some 846 of these cars were delivered with factory air conditioning (designated M573).

Circular side reflectors at front and rear show this Martini limited edition to be to US specification. Standard features on this model were wider 6J × 14 cast alloy wheels and anti-roll bars at front and rear.

28

924

What really set the 1977 Martini 924 models off was the special interior trim with scarlet carpets and seat inlays, plus Martini & Rossi stripes on the headrests and blue piping around the seat edges. Each car had a special centre console plaque commemorating the World Championship of Makes victories.

924 Le Mans (1980)

By 1980, sales of the 924 were slipping and it was decided that another 'special' for Europe was needed to make the range more appealing. That year, three heavily-modified 924 Carrera GTs had been entered at Le Mans, with one achieving the remarkable feat of finishing sixth overall. Not long after, the limited edition 924 Le Mans (again designated M426) was announced in Europe. In total 1030 were built, 100 to UK specification.

These cars were painted Alpine White, with subtle red, yellow and black 'cheat lines' round their waists. Ahead of each door was a stylised 'Le Mans' script, with '924' scripted on the door sills. Adding to the looks and improving high-speed stability was the 924 Turbo's flexible rear spoiler, while the cars were also equipped with a removable sunroof, rear wiper and passenger door mirror.

The seats were black leatherette, with a black and white pinstripe fabric and white piping. There was a new 360mm (14.17in) four-spoke steering wheel with a leather rim, along with a leather sleeve on the gear lever. The new steering wheel might have helped with getting in and out of the car, but, combined with the wide 205/60HR15 tyres on spoke-effect 6J × 15 wheels, the unassisted steering was extremely heavy at low speeds.

The 125bhp engine was standard, along with the new Audi five-speed gearbox, but attention

Original Porsche 924/944/968

was paid to the suspension by uprating the shock absorbers and fitting anti-roll bars front and rear.

The Le Mans is considered desirable because it combines all the refinements of later 924s with a large range of sought-after optional extras.

924 Weissach (1981)

By 1981 life for the 924 in the US was becoming even more difficult. Japanese competition from cheaper models offered by Mazda and Nissan had significantly eaten into market share and yet again the car needed the boost of special models to catch buyers' attention.

The 924 Weissach, named after Porsche's well-known research centre, was just such a limited edition, with only 400 cars made for the US market as factory order M459. The specification was similar to the European Le Mans, but loaded with even more useful options. Externally, all the cars were finished in Pewter (platinum) metallic, with electrically adjustable exterior mirrors. Inside there was an attractive two-tone tweed fabric for the sports seats, as well as electric window lifters and air conditioning as standard.

As with the European Le Mans limited edition, the Weissach is the most desirable of the pre-1981 US models, although it was possible to match its specification by ordering from the regular options list.

Further special 924 models are listed in the data section.

The 1980 924 Le Mans was finished in Alpine White with Germanic red, black and yellow 'cheat' lines along the sides – 1030 were built.

Logo details: the words 'Le Mans' were picked out behind the front wheel arches, and special sill covers carried the '924' motif. Multi-spoke cast alloy wheels on the 924 Le Mans were inspired by those of the Turbo, but differed in having four-stud fixings.

Data Section

IDENTIFICATION DATA

Year	Model	Market	Chassis numbers	Engine type	Gearbox type
1976	924	Eur/RoW	9246100001-5149	XK	YR
1977	924	Eur/RoW	9247100001-8512	XK, XJ	YR
	924	US/Canada	9247230001-5789	XG	XT, RH, RL
	924	Japan	9247300001-0261	XF	XT, RL
	924	US/Can/Jap	9247330001-0164	XG	XT, RL
1978	924	Europe	9248100001-9474	XK	YR, RC
	924	US	9248200001-11638	XG	XT, RH, RL
	924	Japan	9248330001-0450	XF	XT, RL
1979	924	Europe	9249100001-10475	XK	YR, RC
	924	US/Cal	9249200001-09636	XG	XT, RH, RL
	924	Japan	9249330001-0508	XG	XT, RL
1980	924	Eur/RoW	92A0410001-9094	XK	VQ, RK
	924	US/Japan	92A0430001-3700	VC	VR, RL
1981	924	Eur/RoW	WPOZZZ92ZBN400001-9669	XK	MD, RK
	924	US/Japan	WPOAA092BN450001-2155	VC	MF, ME, RL
1982	924	Eur/RoW	WPOZZZ92ZCN400001-7814	XK	MD, RK
	924	US/Japan	WPOAA092CN450001-2277	VC	MF, ME, RL
1983	924	Eur/RoW	WPOZZZ92ZDN400001-5789	XK	MD, RK, RCC
1984	924	Eur/RoW	WPOZZZ92ZEN400001-4659	XK	MD, RCC
1985	924	Eur/RoW	WPOZZZ92ZFN400001-3214	XK	MD, RCC

Notes
The final entry for 1977 is the 110hp SAE '1977.5' model for US/Canada/Japan. The last chassis numbers shown are the last vehicle built for each model year. The factory reserved the first 60 chassis numbers for internal use, so these figures do not necessarily represent the actual numbers built.

PRODUCTION DATA

Year	Model/market	Max power (bhp@rpm)	Max torque (Nm@rpm)	Compression ratio	Weight (kg)	Number built
1976	924 Eur/RoW	125@5800	165@3500	9.3:1	1080	5149
1977	924 Eur/RoW	125@5800	165@3500	9.3:1	1080	8512
	924 US/Canada	95/110@5750	148@4000	8.0/8.5:1	1291	5789
	924 Japan	95/110@5750	151@3500	8.0/8.5:1	1291	425
1978	924 Europe	125@5800	165@3500	9.3:1	1080	9474
	924 US	110@5750	151@3500	8.5:1	1291	11638
	924 Japan	110@5750	151@3500	8.5:1	1291	450
1979	924 Europe	125@5800	165@3500	9.3:1	1130	10475
	924 US/Canada	110@5750	151@3500	8.5:1	1291	9636
	924 Japan	110@5750	151@3500	8.5:1	1291	508
1980	924 Eur/RoW	125@5800	165@3500	9.3:1	1130	9094
	924 US/Japan	110@5750	151@3500	8.5:1	1291	3700
1981	924 Eur/RoW	125@5800	165@3500	9.3:1	1130	9669
	924 US/Japan	110@5750	151@3500	8.5:1	1291	2155
1982	924 Eur/RoW	125@5800	165@3500	9.3:1	1130	7814
	924 US/Japan	110@5750	151@3500	8.5:1	1291	2277
1983	924 Eur/RoW	125@5800	165@3500	9.3:1	1130	5789
1984	924 Eur/RoW	125@5800	165@3500	9.3:1	1130	4659
1985	924 Eur/RoW	125@5800	165@3500	9.3:1	1130	3214
Total						110,427

Notes
Power figures for US models are quoted in hp SAE. The second power figure quoted for two 1977 versions denotes the '1977.5' US/California/Japan model. The 1977 model year production figures for the US should be treated with caution, as every source consulted said something different!

IDENTIFICATION

Chassis numbers Between 1976-79, 924 chassis numbers contained 10 digits, structured as follows: first three digits, model type (924); fourth digit, model year (6, 7, 8 or 9); fifth digit, version (1 = Europe 125bhp, 2= US/California/Canada 90hp, 3= Japan 90hp); last five digits, serial number (00001 onwards). US/Japan '1977.5' models (110hp) were given an additional digit 1 after the version number (ie, 924721 or 924731, etc).

The pre-1980 pattern of chassis numbering, with 10 digits. This is a 1977 model year car (denoted by '7' for the fourth digit) to US specification (denoted by '2' for the fifth digit). This is the additional plate displayed on US models behind the windscreen on the driver's side.

With imminent EEC and US legislation necessitating a change in the system for 1981, an interim 10-character notation was used for 1980, and also carried over to the Carrera GT and Carrera GTS/R. For the 1980 model year only, a 924 chassis number was structured as follows: first two characters, model type (92); third character, model year (A for 1980); fourth character non-descriptive (O); fifth character, vehicle type (4); sixth character, engine type (1 = Europe/RoW 125bhp, 3 = US/Japan 110hp); last four digits, serial number (0001 onwards).

From 1981 chassis numbers were identifiable on a world-wide basis for a period of 30 years by a 17-character system that is still in use, called the Vehicle Identification Number (VIN). A 924 VIN would comprise these characters in sequence: first three letters, world make code (WPO); next three characters, US vehicle specification code (rendered ZZZ for RoW models) – first letter is body type (A = coupé), second letter is engine type (A = 2-litre US), third number is occupant safety system (0 = safety belts only, 1 = driver airbag, 2 = driver/passenger airbags); seventh and eight digits are vehicle type (92); ninth character is a test code (Z) or left blank on US models; 10th character is model year code (B

A chassis number to the post-1981 VIN system, with 17 characters. This car is a 1985 model year car (denoted by the 'F') to 'Rest of World' specification (indicated by the 'ZZZ'). On all cars the chassis number was embossed on to the firewall in the engine compartment.

Original Porsche 924/944/968

Summary of Special Models

Name	Factory order number	Number built	Production date	Colour	Interior
Martini	M426	1000 (RoW) 2000 (US)	12/76 to 03/77	White	Black/red
US	M426	1800	03/78 to 07/78	Dolomite Grey met	Black/silver
Switzerland	–	100	02/78 to 03/78	Pearl met	Brown, pinstripe
France	M427	100	09/78 to 10/78	Dolomite Grey met	Black, pinstripe
Sebring '79	M429	1400 (US)	12/78 to 02/79	Guards Red	Black, red/blue
Cork [1]	M428	300 (RoW)	03/79	Pearl met	Brown, pinstripe
Italy	–	150	02/80	Black met	Black, pinstripe
Le Mans	M426	1030 (RoW)	07/80 to 08/80	White	Black, pinstripe
US	M459	400	03/81 to 04/81	Platinum met	Brown, beige
50th Jubilee	M402	589 (W. Ger) 425 (RoW)	07/81 to 09/81	Pewter met	Black, leather, grey Black, leatherette, grey
Switzerland	M449	30	12/81 to 01/82	White	Black, pinstripe
Italy	M426	100	03/82 to 04/82	Black	Black, pinstripe
W. Germany	I	58	6/82	Montego Black	Black, grey
	II	32	6/82	Sable Brown	Brown, beige
	III	28	6/82	Sable Brown	Beige, brown
	IV	16	6/82	Zermatt Silver	Black, red/blue
	V	22	6/82	Zermatt Silver	Black, green/blue
	VI	12	6/82	Zermatt Silver	Black, grey/black

Note
[1] This model was for non-US export markets only.

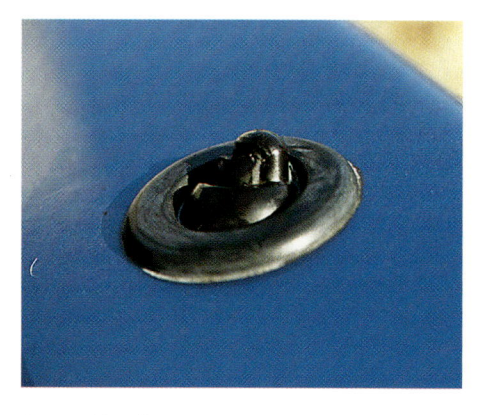

Optional headlamp washer nozzles (above) were fitted to the top surface of the front bumper. A 924 with the sunroof panel removed (below) shows off the wind deflector at the front edge of the opening.

= 1981, C = 1982, etc); N is manufacturing location (N = Neckarsulm); 4 is vehicle type; 5 is body/engine code; and finally the serial number is given (starting from 0001).

The 1979 model year was the last for a specific 924 for the Japanese market. For 1980 and '81, Japanese models were the same as US specification. The 924 was discontinued in US/Japan for 1983.

Engine numbers From 1976 onwards 125bhp RoW engine serial numbers were prefixed XK followed by a six-figure number (eg, XK000001). Right-hand drive RoW engines in 1978/79 were designated XJ. For 1977 '49-state' US/Canada 90hp engines were designated XH, and California/Japan 90hp engines XF. For 1978 California/Japan engines received special emissions equipment, denoted by XE. From 1980 onwards US/Canada/Japan engine serial numbers started VC.

Gearboxes Numbered by type, followed by a five-figure number series denoting date of build (eg, YR19026 – built on 19 February 1976).

The four-speed Audi unit was coded 088/6/YR (Europe/RoW) or 088/A/XT (US/Japan). For 1978 the Getrag five-speed gearbox (016) was optional on the 924, coded 016Z/VA (Europe) or 016Y/VB (US). If the serial number has the digit 12 after the designator then a limited slip differential is fitted.

From 1980 the Audi five-speed gearbox was introduced, of type 016/8 (Europe/RoW, initially with serial number prefix VQ) or 016/9 (US/Canada/Japan, initially with prefix VR). From 1981 the prefixes changed to MD (Europe/RoW), MF (US/Canada) or ME (Japan). When fitted with a limited slip differential, the code was 4Q (Europe/RoW) or 5Q (US/Canada/Japan).

Three-speed automatic transmission was optional for Europe/RoW markets from October 1976, denoted by serial number prefix RC, followed by date of manufacture. Prefix was RK on RHD cars, RH for US/Canada, RL for California/Japan. For 1980, the US/Canada/Japan autos were standardised on the prefix RL (with 11/41 final drive previously used for California/Japan). Europe/RoW autos were standardised on prefix RK (with 11/38 final drive). From October 1982, Europe/RoW auto prefix changed to RCC.

Racers In the 1980 model year 16 SCCA specification models were delivered to the US with the chassis number series 92A0490001 to 92A0490016.

Key Facts

Fuel European version: 98 octane RON. US version: lead-free.
Dimensions Length, 4213mm; width, 1676mm (US version, 1685mm); height, 1270mm; wheelbase, 2400mm; ground clearance, 125mm minimum; turning circle, 10.1m; kerb weight, 1080kg (1976 Euro manual), 1195kg (1977 US auto), 1130kg (1979 Euro).
Capacities Engine oil, 4.5 litres (1976), 5 litres (1979) or 5.5 litres (1984); transmission, 2.6 litres (Audi manuals), 6 litres (auto); fuel tank, 62 litres (1976) or 66 litres (1980); screenwash, 2 litres; headlamp wash, 6.4 litres.

Options

Several markets specified some of these factory options as standard equipment. M numbers are given only where they might be useful for identification, either through an original vehicle invoice or, after 1982, on a chassis plate.

Jan 1976 Removable roof panel; four stereo speakers, manual aerial, wiring, etc; tinted windscreen and side glass; manual aerial wiring for radio; spotlights; convex driver's door mirror; M404 front and rear anti-roll bars; rear fog lamp; oil pressure and temperature gauges; emergency spare wheel with compressor; 63Ah battery; metallic paint. US-only specials. Touring Package 1: 6J x 14in alloy wheels with 185/70HR14 tyres, leather-covered steering wheel. Touring Package 2: rear wiper, headlamp washers, passenger door mirror.
Jan 1977 Air conditioning.
Jun 1977 Three loudspeaker system, electric aerial and wiring for radio; electric aerial only; M474 sports dampers (only with front/rear anti-roll bars); five-speed gearbox; Porsche CR stereo radio (three types: type US stereo cassette; type RW for other markets or type DE for Germany with traffic decoder).
Aug 1977 British Lux option, including alloy wheels with 185/70HR tyres, tinted windows, rear wiper and headlamp washers.
Jan 1978 Electric windows; alloy wheels with black centres; left exterior electric mirror.
Aug 1978 British Lux option expanded to include electric windows and electrically operated driver's door mirror.
Aug 1979 Leather seats; 380mm three-spoke leather steering wheel; 360mm four-spoke leather steering wheel; two-tone paint; 1.1kw alternator, with 63Ah battery for automatics; bumpers with impact absorbers and rubber overriders; electric driver's door mirror; alarm. M602 Special Package 2: special trim (velour), centre console, 924 decals, electric windows, rear fog lamp.
Apr 1980 380mm three-spoke steering wheel with 30mm raised hub; M450 6J x 15in cast multi-spoke road wheels with 205/60HR15 tyres. M468: graduated tint windscreen, tinted side glass, convex left outside electric mirror, mechanically adjustable exterior door mirror (and convex).
Feb 1981 Digital radios; sports seats in leather; sports seats in Berber/leatherette; cassette container and coin box.
Jul 1981 M220 limited slip differential; light alloy wheels with 205/55VR16 tyres; M404 anti-roll bars front and rear; fuel consumption indicator; wheel locks; sport shock absorbers; M666 no preservation coating; tourist delivery; 924 logos on door sill kick plates.
Note For 1982 model year onwards, see 944 options (pages 73-74).

Colours & Interiors

1976 (chart 1100.14)
Standard colours A1 Black, D7 Rallye Yellow, G2 Scarlet, M8 Palm Green, N2 Signal Green, R3 Magnolia, T4 Maroon.

32

Electric window option (above) provided switches for both windows on driver's door; at lower left is the adjusting rocker switch for the electrically operated exterior mirror. Circular side reflectors (below) were required on US cars.

Special colours W6 Titian met, W7 Sapphire met, Y5 Peppermint Ice met (also called Viper green met), Z4 Diamond Silver met.
Upholstery Leatherette (inlays leatherette perforated): black, Gazelle (tan). Cloth inlays: Tartan Dress: Saddle Brown, Black/Yellow, Black/Silver.
Carpet Black, Gazelle.

1977 (chart 1122.14)
Standard colours A1 Black, D7 Rallye Yellow, G2 Scarlet, H7 Brocade Red, N2 Signal Green, R5 Polar White.
Special colours W2 Copper met, W3 Turquoise, Y3 Bahama Blue met, X3 Reseda Green met, Z4 Diamond Silver met.
Upholstery Leatherette (inlays leatherette perforated): Black, Gazelle (tan). Cloth inlays: Tartan Dress: Saddle Brown, Black/Yellow, Black/Silver.
Carpet Black, Gazelle.

1978 (chart 1153.14)
Colours A1 Black, D7 Rallye Yellow, N2 Signal Green, W2 Copper met, G2 Scarlet, P1 Alpine White, X3 Reseda Green met, Z4 Diamond Silver met, H5 Malaga Red, T3 Bitter Chocolate, Y3 Bahama Blue met, Z7 Colibri Green met.
Upholstery Leatherette (inlays leatherette perforated): Black, Tan, Brown. Cloth inlays: Herringbone brown/beige, Herringbone black/white, Herringbone orange/black.
Carpet Beige, brown, black.

1979 (chart 1904.14)
Colours A2 Mocha Black, A5 Lilac, D9 Mexico Beige, G1 Guards Red, H5 Malaga Red, P1 Alpine White, W4 Petrol Blue met, W6 Minerva Blue met, W9 Indian Red met, Z4 Diamond Silver met, Z7 Colibri Green met, Z9 Dolomite Grey met.
Upholstery Leatherette (inlays leatherette perforated): Black, Tan, Brown; leather front and inlay also available in these colours. Cloth inlays: Herringbone brown/beige, Herringbone black/white, Herringbone orange/black.
Carpet Beige, brown, black.

1980 (chart 1005.14)
Standard colours A2 Mocha Black, H5 Malaga Red, A5 Lilac, D9 Mexico Beige, J3 Monaco Blue, K5 Amethyst, G1 Guards Red, G3 Venus Red, P1 Alpine White.
Special colours W4 Petrol Blue met, W6 Minerva Blue met, W9 Indiana Red met, Z2 Onyx met, Z4 Diamond Silver met, Z9 Dolomite Grey met.
Dual-tone colours D9A2 Mexico Beige/Mocha Black, P1D9 Alpine White/Mexico Beige, P1G1 Alpine White/Guards Red, Y4Z2 Inari Silver met/Onyx met, Z4W5 Diamond Silver met/Helios Blue met, Z4Z9 Diamond Silver met/Dolomite Grey met.
Upholstery Leather/leatherette: black, brown, beige. Tartan dress: green/black, red/black, grey/blue/black. Pinstripe: black/white, brown/white. Pasha velour: brown/beige, light brown/cream.
Carpet Beige, brown, black.

1981 (chart number not recorded)
Standard colours A2 Mocha Black, D5 Colorado Beige, G1 Guards Red, G3 Venus Red, J3 Monaco Blue, K9 Mauritius Blue, N8 Conifer Green, P1 Alpine White, T4 Havana Brown.
Special colours U1 Pewter met, U7 Black met, W6 Minerva Blue met, W9 Indiana Red met, Y3 Saturn met, Y5 Meteor met, Z2 Onyx met, Z4 Diamond Silver met.
Dual-tone colours D5A2 Colorado Beige/Mocha Black, U1T4 Pewter met/Havana Brown, Y4Z2 Inari Silver met/Onyx met, Z4Y5 Diamond Silver met/Meteor met.
Upholstery Leather/leatherette: black, brown, beige. Berber cloth: dark brown/white, light brown/white. Pinstripe: black/white, brown/white. Pasha velour: dark brown/beige, light brown/cream.
Carpet Beige, brown, black.

1982 (chart WVK 120614)
Standard colours A2 Mocha Black, A3 Gabon Grey, G1 Guards Red, H2 Gambia Red, K9 Mauritius Blue, P1 Alpine White, T4 Havana Brown.
Special colours U1 Pewter met, U2 Light Blue met, U7 Black met, W1 Ocean Green met, Y2 Claret met, Y5 Meteor met, Z4 Diamond Silver met.
Dual-tone colours P1A3 Alpine White/Gabon Grey, U1T4 Pewter met/Havana Brown, Z4Y5 Diamond Silver met/Meteor met.
Upholstery Leather/leatherette: black, brown, beige. Berber cloth: dark brown/white, light brown/white. Pinstripe: black/white, brown/white. Pasha velour: dark brown/beige, light brown/cream.
Carpet Beige, brown, black.

1983 (chart WVK 100114)
Standard colours A1 Black, B4 Pasadena Yellow, G1 Guards Red, K3 Copenhagen Blue, P1 Alpine White.
Special colours L1 Zermatt Silver met, L2 Gemini Grey met, L3 Montego Black met, L4 Sable Brown met, L5 Sapphire met, R6 Light Bronze met, U1 Pewter met, W9 Siena Red met, Y8 Moose Green met.
Upholstery Leather/leatherette: black, brown, grey-beige. Pasha velour: light grey/black, brown/grey. Pinstripe: black/white, brown/beige, grey-beige/white. Berber cloth: grey/black, beige/brown, grey-beige/brown.
Carpet Black, brown, grey-beige, grey.

1984 (charts VMA7.83, WVK 103120)
Standard colours A1 Black, B4 Pasadena Yellow, G1 Guards Red, K3 Copenhagen Blue.
Special colours L1 Zermatt Silver met, L2 Gemini Grey met, L3 Montego Black met, L4 Sable Brown met, L5 Sapphire met, R6 Light Bronze met, U1 Pewter met, X7 Ruby Red met.
Upholstery Leather/leatherette: black, brown, grey-beige. Pasha velour: light grey/black, brown/grey. Pinstripe: black/white, brown/beige, grey-beige/white. 'Porsche' cloth: black, brown, grey-beige.
Carpet Black, brown, grey-beige, grey.

1985 (charts VMA10.84, WVK 103221)
Standard colours A1 Black, D4 Pastel Beige G1 Guards Red, K3 Copenhagen Blue, P1 Alpine White.
Special colours L1 Zermatt Silver met, L5 Sapphire met, S2 Garnet Red met, S5 Crystal Green met, U8 Stone Grey met, W9 Graphite met, Y4 Kalahari met, Z6 Mahogany met.
Upholstery Leather/leatherette: black, brown, burgundy, light grey. Pinstripe: black/white, brown/beige, burgundy/white, light grey/white. Pinstripe flannel: anthracite, brown, burgundy, light grey. 'Porsche' cloth: black, brown, burgundy, light grey.
Carpet Black, brown, burgundy, light grey.
Special leather/carpet interiors Pearl White, Buff Skin Brown, Can Can Red, Champagne, White.

Pasha fabric – guaranteed to induce dizziness!

Original Porsche 924/944/968

The 924 Turbo (1979-83)

From its inception, one of the greatest criticisms levelled at the 924 was its distinctly average performance. Given Porsche's experience of turbocharging, first with its CanAm sports racing cars of the early '70s and then with the 911 Turbo, it was an obvious progression to apply the same technology to the 924.

The 911 Turbo had been a bold stroke, authorised by the same board meeting in 1974 which had also made the far-reaching decision to buy the 924 from VW-Audi. At this time, after the world oil crisis in 1973, the idea of producing such an exclusive, and expensive, luxury grand tourer cut right across the market trend. But Porsche had

Evolution Outline

Aug 1978	Start of production of 170bhp model.
Aug 1979	Start of deliveries of US 143hp (SAE) model. Flap on fuel filler.
Aug 1980	177bhp with Digital Motor Electronics (US model improves to 154hp). Fuel tank enlarged to 84 litres. 400 Carrera GTs built. Body guarantee extended to seven years.
Jan 1982	Run-out of US models.
Jun 1982	Run-out of Rest of World models.
Dec 1983	Production ceases after build of Italian models.

34

924 Turbo

This 1980 model Series 1 924 Turbo displays the extra openings that appeared in the bodyshell to keep the brakes and engine cool.

As seen on the facing page, the addition of a rear spoiler helped set the 924 Turbo apart from other 924s. This car is finished in one of the attractive metallic paints – Diamond Silver in this case – that always helped the 924 Turbos to look more purposeful.

been rewarded by the 911 Turbo selling in large numbers for such a specialist model.

Development of the 924 Turbo was begun soon after it became clear the 911 Turbo was going to be a success. Porsche believed it could inject sufficient performance into the 924 to generate new enthusiasm for the car, which at this stage was still considered to be vital for the future due to the conviction that the 911 had only a limited life.

The company also hoped the 924 Turbo would help overcome the severe image problem the regular 924 was already suffering. The core group of long-time Porsche enthusiasts remained unimpressed with the four-cylinder car's lack of charisma, despite its record-breaking sales. All the company's racing success, they would point out, had been achieved using either sports racing or 911-derived models. The 924 had not become the competition vehicle of anybody's choice, even in club racing.

The 924 Turbo, it was therefore argued, would help the car shrug off its VW-Audi parentage to become adopted as a true member of the Porsche family. The engines were to be assembled by Porsche at Zuffenhausen, with the 'short' blocks shipped in from Salzgitter and then sent on to the 924 production line at Neckarsulm.

To further impress die-hard grumblers, the external changes compared to the regular 924 were deliberately distinctive. From any angle the car stood apart. There was no question, either, that the performance changes had made all the difference, transforming the car to such an extent it seemed its well-balanced chassis would be capable of handling still more power.

The new model, with internal type number 931, was eventually launched in the spring of 1978 and went into production for the 1979 model year. It boasted 170bhp in European form, some 36% more than the unblown version. As with the 911 Turbo which had preceded it, the 924 Turbo was much more than just a turbocharged standard car. The chassis had been given a thorough work-over to improve its dynamics, while the braking system had been boosted to match the higher engine output.

US versions were not shipped until after the start of the 1980 model year, when the car became

Original Porsche 924/944/968

the only turbocharged Porsche available there. The 911 Turbo had already been withdrawn after turning out to be too costly to convert to the new exhaust emissions standards. All 924 Turbos sold in the US were '50-State' cars, losing a lot of power through being equipped to meet Federal standards. Yet they still gave out 143hp (SAE), 23% more than the regular 924. It was at this time that US models were also fitted with the infamously inefficient sealed beam headlamps, in contrast to the halogens for other markets, along with the equally infamous 85mph speedometers.

From the start the Turbo benefited from the 924's development programme in its early production years. The power to weight ratio of the 170bhp cars was also getting close to 911 territory. For 1981 a 177bhp model (described in this book as the Series 2) was made available in most markets. As is so often the case with Porsche, there was more to this car than met the eye. For the first time, Digital Motor Electronics brought engine management to a genuine production model, improving power, flexibility and fuel consumption. More importantly, improved crankcase breathing was introduced to ensure turbo reliability, a concern with the early Series 1.

Yet, in the eyes of Porsche's all-important accountants, the 924 Turbo was not a success. It failed to make the breakthrough with enthusiast customers and sales stubbornly remained at about 10% of the regular model. In the US, the 924 Turbo was replaced by the 944 in 1982, but a 2-litre tax break in Italy ensured the car's continued popularity there through to 1983.

The 924 Carrera GT, a further attempt to woo the enthusiasts, was built as the result of an order to the competitions department to develop a racing version of the 924. After a prototype was shown, three were raced at Le Mans in 1980. In parallel, 400 productionised road models were built in the 1980 model year, making the Carrera GT the link between the 924 and the 944. In both historical and performance terms, the Carrera GT is a classic Porsche.

The 924 Turbo, particularly the Series 2 model, remains one of the best choices for any enthusiast new to Porsche who is looking for maximum fun for the money. The kick in the back from the turbo aroused some criticism when the car was new, but these days it is part of the high 'grin factor' of what has become a classic among early water-cooled Porsches.

Bodyshell

The 924 Turbos were assembled on the same Neckarsulm production line, but were quite distinct in appearance from the regular 924.

The 924 Turbo was almost as quick as the contemporary 911SC, both in terms of acceleration and top speed. This smart 1981 model is a Series 2 – European Series 2 versions are easily identified by the small indicator repeaters on the front wings.

924 Turbo

Four openings in the front panel made sure those ahead knew this was no ordinary 924. The elegant NACA duct in the bonnet sucked in cooling air over the turbocharger when moving, and allowed hot air to exhaust from the engine compartment when at standstill.

At the front the new model featured four cooling air intakes just above the bumper and between the headlamps, along with further grilles on each side of the front spoiler to improve cooling to the brakes. A NACA-shaped duct on the bonnet drew cooling air over the turbo when the car was moving, and provided an escape path for hot air when the car was stationary. An undertray helped to create a low pressure area beneath the engine compartment which drew hot air out at speed. At the rear a new black polyurethane spoiler not only reduced drag, but also significantly improved appearance.

As making the Turbo look different was considered so important, modifications were also carried over to the paint finish. The two-tone colours were racy and these days add a period touch. The colours split along a horizontal line drawn from the lower edge of the rear light cluster to the top edge of the front bumper. There is no question that those which have best withstood the test of time are the darker metallics, particularly Claret, Pewter and Meteor.

With the 1981 models the long-term guarantee against rust perforation was extended to seven years. From the start the whole bodyshell had been manufactured from zinc-coated steel.

The fuel tank initially remained at 62 litres (13.6 Imperial gallons, 16.4 US gallons), but with

Original Porsche 924/944/968

an extra fuel pump fitted inside the tank itself, along with larger fuel lines. The tank was increased to 66 litres (14.5 Imperial gallons, 17.4 US gallons) for the 1980 model year, and for the 1981 model enlarged yet again to 84 litres (18.5 Imperial gallons, 22.2 US gallons) and equipped with a new sender unit.

Body Trim & Fittings

The equipped weight of the Turbo was about 100kg (220lb) more than the first 924s and about 50kg (110lb) more than the 1979 model. Of this 29kg (64lb) was due to the turbo installation, strengthened exhaust, inlet ducting and the pipework for the oil, etc.

The new rear spoiler gave an undoubted performance advantage by reducing the drag coefficient from 0.36 to 0.33. But it also gave owners a useful hand-hold with which to close the rear glass, which led to an unexpected problem. Because of the extra leverage many owners got into the habit of closing the glass with one hand, at the same time usually pulling to one side, so the early gas struts quickly wore and the sealing around the rim of the glass became ineffective. Caring owners should always use two hands.

A new 'turbo' decal appeared on the rear panel under the spoiler, with the '924' it replaced on early cars being moved to the left side. Apart from black anodised trim for the

Early 924 Turbos used a decal (left) for the rear designation, but for the 1981 Series 2 models the appearance was improved with a proper badge (centre) of the style found on the 911 Turbo. The magic word was repeated on the door sill trims (right).

924 Turbo

side window, external trim was otherwise as the corresponding 924 1980 model year.

1981 Series 2 models are easily identified by the side repeaters introduced on each front wing.

Interior Trim

Continuing the effort to make the Turbo stand out from the crowd, the interior options were increased to include the new Herringbone trim for 1979. For the US market, the chequered Pasha, an early '80s Porsche favourite, was introduced across the range.

At the car's launch a link was made with the 911 Turbo by using the same design of 380mm (15.0in) diameter, three-spoke, leather-covered steering wheel. The 1981 model year then saw a new four-spoke design by the Scottish stylist Dawson Sellar, who was then working for the company. At 360mm (14.2in) diameter, this was 20mm (0.8in) smaller than the wheel of the regular 924.

Dashboard & Instruments

The engineers did not consider a boost gauge necessary, although buyers maintained that one would have been nice. Meanwhile owners concerned with turbocharger life could replace the small clock in the centre console with an oil temperature gauge, which at least gave some indication whether the hard-worked engine oil was getting too hot.

They were also given a new 260kph speedometer (170mph for the US and the UK), while the instrument lettering was changed from the white of the regular model to green. This 'improvement' was scrapped with the 1981 model, which reverted to the more easily read white.

Luggage Compartment

There were no changes to the regular 924's highly successful luggage area.

Engine

As the Turbo's engine was built in Zuffenhausen it can safely be called 'real' Porsche. 'Short' blocks – consisting of crankcase, crankshaft, forged pistons and connecting rods – were shipped in from Salzgitter. A brand new cylinder head was fitted and completed engines were test run before going to Neckarsulm for installation.

The new aluminium-silicon alloy head still comprised two in-line valves per cylinder, but the combustion chamber was now dished, preventing premature detonation and allowing longer valve

Interior contrasts. Pasha chequered upholstery (facing page) was a very popular choice for the interior. Even the lighter shades (below) can look good after many years of ownership if well looked after.

Original Porsche 924/944/968

travel. A new cylinder head gasket, supplied by the Reinz company, was designed to cope with the much higher operating temperatures of a turbocharged engine by featuring special combustion chamber seals, offering low distortion and requiring no re-tightening. The exhaust valves were increased in diameter by 3mm (0.12in) to 36mm (1.42in), with the inlets remaining the same at 43mm (1.69in). The drive train to the valves was unchanged. The redesign also moved the platinum-tipped spark plugs to the induction side (and closer to the inlet valves), with the fuel injectors now screwed into the head. Compression ratio was reduced from 9.3:1 to 7.5:1. The new pistons, either iron or chromium coated depending on whether they had been supplied by Mahle or Karl Schmidt, both had the same shape to the crown.

The KKK (Kuhnle, Kopp & Kausch) K26 turbocharger was a truck unit, fitted downstream of the exhaust manifold under the right side of the engine, bumping the starter motor over to the left. Although the turbo was sized to spin up to its maximum speed of 100,000rpm at engine revolutions as low as 2750rpm, there was a noticeable delay after pressing the throttle before boosted acceleration made itself felt.

Maximum charge pressure was 0.7 bar except on US models, which had a 0.45 bar limit determined by the need to prevent the catalysed engine knocking on lead-free fuel. Over-pressure in the mid- to high-revolution ranges was controlled by an exhaust by-pass valve, with a pop-off valve controlling pressure when the throttle was closed. There was no induction manifold intercooler for the inlet air charge. A cast duct tracked over the top of the cylinder head to the inlet throttle body and plenum.

The turbocharger's bearings were lubricated by engine oil and keeping the temperature of this low was a major design problem. Unlike the 911 Turbos, whose engines were less stressed and more likely to be used on longer runs, 924 Turbos were practical everyday cars which were subjected to continuous stop-start lifestyles. Turbo reliability became a big issue, with smoky exhausts soon a regular sight after turbo bearings had quickly worn out. Using a high-quality oil and allowing the oil to cool after a run – by letting the engine idle for a few minutes – were critical to prolonging bearing life. Otherwise there was a tendency for the oil to boil – not conducive to good lubrication of the turbo's bearings. An external oil cooler, mounted just ahead of the water radiator, indicated the severity of the problems encountered during the engine's development.

This four-spoke steering wheel was the most obvious distinguishing feature of the Series 2 924 Turbo interior, but note also the 260kph speedometer. This car has air conditioning, with the control on the centre console in the middle instrument location. The clock has moved to the right-hand side, replacing the voltmeter, and the small control in front of the radio is the speaker fader.

40

924 Turbo

The main changes for the Series 2 924 Turbo (from the 1981 model year) were in the engine compartment. Improved crankcase breathing went some way to promoting oil recirculation in the turbocharger after shut-down, but careful ownership was still the answer for long turbo life.

The Digital Motor Electronics (DME) system used on the Series 2 924 Turbo was the first step towards engine management. This is the distributor, mounted off the rear of the camshaft.

A turbocharged engine needs closer control of the ignition than a normally aspirated one because of the larger variation of charge mass and inlet pressure and temperature. On the first Turbos the 924's existing transistor system was used with centrifugal advance, and the Bosch K-Jetronic system was adjusted to supply the increased fuel requirement. In conjunction with the main pump, fuel delivery pressure now became 170 litres/hour at 6 bar pressure, which was enough to overcome most problems with fuel vaporisation. Further fuel supply was simply cut off if turbo boost exceeded 1.1 bar.

The result of all this was that the engine compartment suddenly becoming extremely crowded, to the extent that oil filter changes were now carried out from underneath. Thankfully, though, the new position of the spark plugs – on the induction side – made them easily accessible because of the tilt of the engine.

The engine for most markets delivered 170bhp at 5500rpm with maximum torque of 244Nm (180lb ft) at 3500rpm. The American version, loaded with emission control equipment, still offered 40% more power than the regular US 924, with 143hp (SAE) at 5500rpm and maximum torque of 199Nm (147lb ft) at 3000rpm. The slightly smaller US turbocharger – to give better mid-range torque – actually improved catalytic converter life, as it shielded the core from the intense heat of the exhaust manifold.

The engine was significantly revised for the 1981 model year by adopting Siemens-Hartig digital ignition, known as Digital Motor Electronics (DME). This was the first engine management computer ever fitted to a series production Porsche. Its use required revising the flywheel to include a speed sensor at its rim, as well as replacing the previous twin throttles in the intake manifold by a single large throttle. Sensors then monitored air inlet temperature, throttle butterfly position and inlet suction pressure to optimise fuel delivery and timing. As well as usefully improving power and torque, the system significantly improved fuel consumption. Other changes involved making the fuel injector bodies in metal rather than plastic and revising the pistons to give a larger recess.

Combined with the close control of the timing – and hence detonation – given by DME, the compression ratio was increased to 8.5:1 for non-US models, while charge pressure was eased to 0.64 bar. In all, this raised power to 177bhp and maximum torque to 250Nm (185lb ft). For US models required to run on lead-free gasoline, the compression ratio was raised to 8.0:1, with boost eased to 0.44 bar. This nonetheless increased the power of US models to 154hp (SAE) at 5500rpm, with maximum torque rising slightly to 210Nm (155lb ft) at 3300rpm.

These improvements were not so noticeable in overall performance, but they did make the engine much more flexible. The same turbos were also now being used for all markets after much attention had been paid to improving crankcase breathing so that the oil could circulate naturally after the engine had been shut down.

Transmission

The new Turbo's power was transmitted through a larger 225mm (8.86in) diameter single dry-plate clutch, operated hydraulically. The torque shaft to the rear transaxle was increased in size from 20mm (0.79in) to 25mm (0.98in), and its different natural frequency of vibration now meant it ran in three bearings rather than four.

After the novelty of the Turbo's extra power had worn off, what impressed most about the car was the new five-speed gearbox. This was a Porsche-developed Getrag 'box with a racing shift pattern that was awkward in slow traffic as first was out on its own to the left, with the normal H pattern covering second to fifth. This arrangement, however, was perfect for fast touring, while, thanks to the well-proven Porsche synchronising system, gear selection was now more precise than with the earlier four-speed Audi unit. Rear axle ratio was 4.125:1 (8/33) or 4.714:1 (7/33) for the US model. A 40% locking limited slip differential became a useful option.

For 1981 US models benefited from a more conventional shift pattern for the five speeds when they adopted the Audi 016G transmission, as opposed to the 016/9 used in regular 924s. This transmission was also to form the basis for the upcoming 944.

Electrical Equipment

There were no major differences in the non-engine electrical equipment and accessories from the regular 924 of equivalent model years. From launch electric window lifters and rear fog lights were not fitted as standard by the factory, although many markets specified them. For 1980 they were included in the factory equipment list.

Suspension & Steering

The extra 50kg (110lb) or so the Turbo carried over the regular 924 resulted in stiffer front springs – identified by a blue paint code – and an increase in front anti-roll bar diameter to 23mm (0.91in). When air conditioning was included, the front springs were made even stiffer, but still carried the blue code.

The rear torsion bars were of 23.5mm (0.93in) diameter and the standard anti-roll bar was softened to 14mm (0.55in) as the five-speed gearbox was lighter than the four-speed. Spacers of 21mm (0.83in) width were fitted to each rear wheel hub, increasing the Turbo's rear track to 1392mm (54.8in) so that wider wheels could be accommodated. Firmer shock absorbers were used all round, Konis at the front and Bilsteins at the rear, along with strengthened wheel bearings. Front/rear weight distribution was 49%/51% unladen, and 44%/56% when carrying fuel and two occupants.

American versions were less stiffly suspended, using a 21mm (0.83in) anti-roll bar at the front, with one at the rear optional.

For all markets the steering rack ratio was higher than the regular 924 at 20:1 rather than 19:1 in order overcome the extra effort needed to turn the smaller steering wheel at low speeds.

924 Turbo

The 924 Turbo wore new multi-spoke wheels with five-stud fixings – they looked great but were very difficult to clean. Disc brakes were ventilated, and those on the front came from the 911.

The Carrera GT changed the character of the 924 completely and inspired the styling of the future 944. In some markets the model was known as the Turbo Carrera.

Brakes

The brakes received a major upgrade. Braking was via a diagonally split dual-circuit system, with most markets using ventilated discs front and rear. Those at the front, with a diameter of 282mm (11.1in), were taken from the 911, while those at the rear, with a diameter of 289mm (11.4in), came from the 928, as did all the calipers. The circuit was boosted by a 229mm (9.0in) vacuum servo.

Sadly, the first US Turbos were not given this braking upgrade and came with the standard car's solid 257mm (10.1in) diameter front discs and 230mm (9.1in) diameter rear drums. However, the attractive US 'Sport' package, which many customers chose, included the ventilated disc brake arrangement, along with 16in wheels.

Wheels & Tyres

The new Turbo was given pressure-cast multi-spoke alloy wheels with five stud fixings, the hubs at the rear containing the handbrake drum, as on the 911. Wheel size was 6J × 15in, fitted with 185/70VR15 Pirelli CN36s. Optionally available from the start of the 1980 model year were 6J × 16in wheels, allowing fitting of the new 205/55VR16 low-profile Pirelli P7s.

924 Turbo USA (1979)

This limited edition series was made for the US in June/July 1979. The cars were essentially cosmetic specials, finished in Dolomite Grey metallic, with black leatherette interiors and black/white tartan check inlays to the seats. A total of 600 were manufactured to factory order number 420.

924 Carrera GT (1980)

The Carrera GT was first shown as a prototype in September 1979. Developed from the then-new 170bhp 924 Turbo, it was an attempt to take the image of the water-cooled four-cylinder cars even further upmarket.

The main thrust of this effort was a competition programme the company's managers hoped would put the 924 up among the 911s – or more accurately the 935s – on the track. The engineers, however, knew from the beginning that this was a futile dream, as a feature in the house magazine *Christophorus* conceded in June 1980. Even with 320bhp – nearly double the output of a production Turbo – a Carrera GT could only achieve

Original Porsche 924/944/968

175mph on the Mulsanne straight at Le Mans, some 50mph slower than a contemporary 935. Yet, despite this sobering fact, and against all odds, a Carrera still finished sixth overall.

The prototype needed more development and it was June 1980 before a limited edition of 400, the number required to homologate the car into the FISA Group 3 production sports class of the time, was offered for sale. As befits a Porsche carrying the title of Carrera, its performance matched the contemporary 911, but at a notably lower price.

The 2-litre engine, designated M31/50, developed 210bhp at 6000rpm, with maximum torque of 275Nm (203lb ft) at 3500rpm. The modifications over the 1981 model Series 2 Turbo were extensive. There were lighter forged pistons and a larger turbocharger, which was actually a larger turbine with the same compressor. Boost pressure was set at 0.75 bar, as opposed to the Turbo's 0.65. A large air-to-air intercooler was placed over the top of the engine, with an inlet which made a distinctive opening in the bonnet, improving volumetric efficiency – the amount of air it drew in – by reducing the charge air temperature by some 45°C.

Camshaft timing remained the same, but the camshaft lobes were specially hardened, while the buckets were made of a different material. The oil cooler was moved ahead of the water radiator, while the exhaust system was lightened and suspended from two points rather than three. The clutch pressure plate was identical to the 911. The gearbox was the Porsche/Getrag G31/03 with third, fourth and fifth gears shot-blasted to improve durability, while first gear synchronisation was taken from the 911, giving a better change from cold.

The front suspension was improved by fitting harder joint bushings and the front ride height reduced by 10mm (0.39in) through shortening the firmer coil springs, which had a red paint marking and an uncompressed length of 359mm (14.13in). The 23mm (0.91in) front anti-roll bar was taken from the Turbo, but the rear bar was stiffened from 14mm (0.55in) to 16mm (0.63in). The rear spring strut angle was reduced from the 18° 5' of the European models to 15° 30', which lowered the rear by 15mm (0.59in). The 21mm (0.83in) spacers remained the same, but the rear trailing arms were stronger around the wheel bearing housing and shock absorber mount, with the drive shafts made of better wearing material.

Forged aluminium Fuchs 7J × 15 wheels were the natural choice for a Porsche with this kind of performance. The offset (installed depth) was reduced to 23.3mm (0.92in), the same as the standard 911, as opposed to the 53mm (2.09in) of the normal 924 Turbo. In conjunction with the 215/60VR15 tyres, this gave a wider track. Further options were 7J × 16 and 8J × 16 wheels, fitted with either 205/55 or 225/50VR16 tyres.

The change in the front wheel offset altered the previously negative scrub geometry to positive, affecting high-speed steering response and prompting a change to a stepped tandem brake master cylinder. This differed from the normal

Tail badging introduced a fresh graphic style – a decal with letters in black outline. The pretty NACA bonnet duct of the Turbo was replaced by a hungrier air scoop. The Carrera signature on the top of the plastic right-hand front wing was a no-cost deletion option from the factory.

The air-to-air intercooler of the Carrera GT put space under the bonnet at a premium.

Interior trim for the Carrera GT and GTS was largely the same as on other 924s, but sculpted sports seats were a welcome addition.

conventional tandem master cylinder by preventing all the force going to the lightly loaded rear wheels in the case of failure of the front system, or during panic braking.

But it was the change in the Carrera GT's appearance which was the most notable feature and overshadowed the other very considerable modifications. The work of Anatole 'Tony' Lapine, then head of Porsche's styling department, it was the prototype for the forthcoming 944's aggressive look. Flexible polyurethane was used for the flared front and rear wheel arches and the deep front air dam, which contained special ducts behind the narrow inlet to feed air to the front brakes. The rear spoiler was significantly larger than on the standard Turbo. The side repeaters on the front wings were made pencil-thin to reflect the car's squatter look.

The only colours were Black, Guards Red or Diamond Silver. The interior was like the Turbo's, featuring black leatherette and black/red pinstripe velour inlays to the seats.

Weight came out at 1180kg (2601lb), only 23kg (51lb) less than the Turbo, but the extra power made all the difference. The 0–60mph acceleration time was cut from 7.5sec to 6.7sec and there was no question of the bulbous wheel arch flares cramping top speed, which was 150mph compared to the Turbo's 143mph. Everyone could now see that the 924 could cope with a great deal more power.

The Carrera GT was a Europe-only model, with 200 sold in Germany, 75 right-hand drive models going to Britain, and the rest of the production batch of 400 being distributed around France, Austria, Italy and Switzerland. Six additional cars were built for either factory development or public relations.

The car was also the only four-cylinder water-cooled Porsche to receive the 'Carrera' title, traditionally awarded only to Porsche's highest performance models, although it was to become cheapened by widespread use later. The 924 Carrera GT fully deserves its classic status.

924 Carrera GTS (1981)

The GTS and GTR were both built in the competitions department at Weissach and could only be bought direct from the factory.

The GTR came before the GTS, by virtue of the four prototype racers that ran at Le Mans in 1980. These competed in the GTP class, with virtually no hope of even a class victory against

Original Porsche 924/944/968

the other highly developed 'silhouette' racers. For the car to get into the Group 4 class, FISA required a production run of 50. Eventually 59 GTS models were built – including those converted to hotter Club Sport specification – along with, it is believed, another 17 GTR racers with 375bhp engines.

The GTS weighed 1121kg (2473lb), approximately 60kg (132lb) less than the standard Carrera GT. All were left-hand drive and finished in Guards Red. They developed 245bhp at 6250rpm, with maximum torque of 335Nm (247lb ft) at 3000rpm. This power increase over the standard GT was achieved by 1.0 bar boost pressure allied to a compression ratio of 8.0:1. While the distinctive scoop on the bonnet was

The 924 Carrera GTS was one step removed from a racing car for the road. Five times Le Mans winner Derek Bell has owned this car from new.

924 Turbo

The Carrera GTS dashboard was unchanged in design, but more instruments were packed in. The rev counter, now placed in the centre, has an inset boost gauge, and an oil temperature gauge has replaced the clock on the centre console.

retained, the air-to-air intercooler was enlarged and moved in front of the water radiator, while the clutch was strengthened and a 40% limited slip differential fitted.

Glass-fibre was used extensively for the body panels and all sound-deadening material was removed, resulting in noise echoing loudly round the interior from stones and grit thrown up by the tyres. The competition background was also evident in the basic interior, which had deep competition-style seats, inertia reel full harness seat belts and doors without side pockets. Yet, as this car was a racer converted for the road, there was a heater/demister and multi-speed wipers.

The firm suspension, on Bilstein gas shock absorbers, featured cast rear trailing arms and coil springs, rather than torsion bars. The cross-drilled ventilated disc brakes were taken from the 3.3 911 Turbo and the forged alloy Fuchs wheels were 7J × 16 (front) and 8J × 16 (rear), with 205/55VR and 225/50VR tyres respectively.

Porsche claimed figures of 0-60mph in 6.2sec and a top speed of 155mph. For brief outline specifications of the non-road legal GTS Club Sport and the GTR models, see the data section on page 48 and the performance table on page 127.

The ventilated discs on the GTS were cross-drilled for better cooling. The Fuchs forged alloy wheels were 7J and 8J.

924 Turbo Italy (1983)

At the very end of Turbo production, in September 1983, a total of 88 special models were manufactured for Italy. They were all finished in Zermatt Silver metallic, with black leatherette and sports seats finished in the new 'Porsche' logo cloth of Anthracite and Burgundy red.

Original Porsche 924/944/968

Data Section

IDENTIFICATION

Chassis numbers For a full explanation see the 924 chapter (pages 31-32). For 1979-80 924 Turbo chassis were given the version number 4 for the fifth digit (for those cars made for Europe/RoW) or 5 (for cars made for US/Japan), in order to identify them from regular 924 models. No 924 Turbos were delivered to the US for the 1979 model year.

Gearboxes For 1978-80 the cars were fitted with Porsche/Getrag G31/01 (Europe/RoW) or G31/02 (US/Japan). From 1981, the US/Japan 924 Turbo Series 2 used the Audi 016G (coded MB for the US, MX for Japan).

KEY FACTS

Fuel 98 RON.
Dimensions Length, 4212mm; width, 1685mm; height, 1270mm; ground clearance, 120mm; turning circle, 10.3m; weight (kerb), 1180kg.
Capacities Engine oil, 5.5 litres; gearbox (manual), 2.6 litres; fuel tank, 62 litres or 84 litres (for 1981 onwards); screenwash, 6.4 litres.

OPTIONS

M numbers are only shown where they might be useful for identification in original vehicle invoices or, after 1982, on chassis plates.

Aug 1979 Leather seats; metallic paint; two-tone paint; two-tone metallic paint; M058 bumpers with impact absorbers and rubber overriders; M220 limited slip differential (40%); electrically operated driver's door mirror; electrically operated passenger door mirror (plain or convex glass); electric windows; headlamp washers; rear wiper; 360mm four-spoke leather steering wheel (black or brown); two extra stereo loudspeakers, electric aerial and wiring; two extra stereo loudspeakers in doors, one in centre, manual aerial and wiring; electric aerial; M474 Sports shock absorbers; M533 alarm; removable sunroof; tinted windscreen and side glass; Air conditioning; 63Ah battery; M351 Porsche CR type DE stereo radio/cassette (Germany), three speakers, manual aerial; M422 Porsche CR type RW stereo radio/cassette (export), three speakers, manual aerial.
Apr 1980 M018 three-spoke leather-covered steering wheel with 30mm raised hub, black or brown; M400 cast alloy 6J x 16in wheels with 205/55VR16 tyres; M468 graduated tint windscreen with side tint glass; mechanical passenger door mirror, plain or convex.
Feb 1981 Digital Radio Super Arimat, aerial, three loudspeakers, etc; digital radio, aerial, three loudspeakers, etc; M409 Sports seats in leather; M410 Sports seats in Berber/leatherette; cassette container and coin box.
1982 onwards See 944 option list (pages 73-74).

COLOURS & INTERIORS

See 924 section (page 33).

IDENTIFICATION DATA

Year	Model	Market	Chassis numbers	Engine type	Gearbox type
1979	924T	Eur/RoW	9249400001-1982	M31/01	G31/01
1980	924T	Eur/RoW	93AO140001-1803	M31/01	G31/01
	924T	US/Can	93AO140001-1803	M31/01	G31/01
		Japan	93AO150001-3440	M31/01	G31/02
	CGT	Prototypes	93BN700001-00006	M31/50	G31/03
	CGT	All	93BN700051-0450	M31/50	G3103
	CGTLM	–	924001-004	924GTP	937/50
1981	924T	Eur/RoW	WPOZZZ93ZBN100001-1783	M31/03	G31/01
	924T	US	WPOZZZ93ZBN100001-1783	M31/03	G31/01
		Japan	WPOAA093_BN150001-1529	M31/04	016G
1981	CGTS	All	WPOZZZ93ZBS710001-0059	M31/60	937/50
	CGTR	All	WPOZZZ93ZBS720001-0012	M31/70	937/50
			WPOZZZ93ZBS720014-0017	M31/70	937/50
			WPOZZZ93ZBS72022	M31/70	937/50
1982	924T	Eur/RoW	WPOZZZ93ZCN100001-0943	M31/03	G31/01
	924T	US/Can	WPOAA093_CN150001-0876	M31/04	G31/02
1983	924T	Italy	WPOZZZ93ZDN100001-0310	M31/03	G31/01

PRODUCTION DATA

Year	Model/market	Max power (bhp@rpm)	Max torque (Nm@rpm)	Compression ratio	Weight (kg)	Number built
1979	924T RoW	170@5500	245@3500	7.5:1	1180	1982
1980	924T RoW	170@5500	245@3500	7.5:1	1180	1803
	924T US/Can/Jap	143@5500[1]	199@3000	7.5:1	1200	3440
	924CGT RoW	210@6000	280@3500	8.5:1	1180	400
	924CGT LM	320@6500	383@4500	7.1:1	930	4
1981	924T Eur/RoW	177@5500	251@3500	8.5:1	1180	1783
	924T US/Japan	154@5750[1]	210@3300	8.0:1	1200	1529
	924CGTS Eur/RoW	245@6250	335@3000	8.0:1	1121	59
	924CGTR Eur/RoW	375@6500	405@5600	7.1:1	945	17
1982	924T Eur/RoW	177@5500	251@3500	8.5:1	1180	943
	924T US/Can	154@5750[1]	210@3300	8.0:1	1200	876
1983	924T Italy	177@5500	251@3500	8.5:1	1180	310
Total						13,351

Note
[1] Refers to hp SAE.

The chassis plate from a Carrera GTS. Normally the 11th character on a 924 would be N (for Neckarsulm), but these cars have an S (for Stuttgart) because they were built in the competitions department at Weissach. As required by German regulations, other information concerns maximum gross weight (1450kg) and maximum axle weights at front (700kg) and rear (800kg).

The 924S (1985-88)

The 2-litre 924 had remained in production, principally for European markets, until 1985, with sales respectable enough to suggest there was still a measurable demand for the clean lines of the original shape, compared to the more beefed-up new 944. But while Porsche was happy enough to keep the 924 turning over, by 1984 a problem was looming over engine supply. The Audi/VW group no longer used the cast iron blocks, and the volume Porsche required – 5789 for the 1983 model year – was not enough for the supplier to remain interested, especially as continued criticism of the engine's heritage and performance meant this figure was expected to fall steadily in coming years. Automobile design was progressing at such a rate in the mid-'80s that over a period of only three years the 2-litre engine had begun to look extremely dated, not only obviously harsh compared to its modern competition, but with higher fuel consumption and exhaust emissions.

But rather than drop the car completely, Porsche decided to give its 'classic' shape a new lease of life by dropping in the smooth-running 2.5-litre engine which had already catapulted the

Evolution Outline

Aug 1985	150bhp 924S introduced for non-US markets.
Jun 1986	US model introduced with same power as other markets.
Sep 1986	Right-hand drive version introduced for the UK.
Aug 1987	Uprated to 160bhp.
Jun 1988	Discontinued.

Without the wide body of the 944, the 924S was an attractive proposition for certain types of club competition. A drag coefficient of 0.33 compared with the 944's 0.35 allowed a 6mph top speed advantage despite having 10bhp less.

Original Porsche 924/944/968

944 to the top of the sports car league. The idea seemed sound, as there was spare capacity on the Neckarsulm production line, while research had indicated that a sector of the market could be reinvigorated by a model like this.

In fact this plan took the wheel the full circle. This was the car the engineers had originally envisaged during the development of the 2.5-litre engine in the late '70s. It had also not escaped Porsche's attention that fitting the 2.5-litre engine into the 924's slimmer bodyshell, meanwhile adding the 924 Turbo's rear spoiler, was likely to produce a faster car. The Turbo's drag coefficient was already 0.33, compared to the 0.35 of the wider 944, while a 2.5-litre 924 would also be slightly lighter.

However, marketing being all, the 924 – as the junior model in the range – could not be allowed to eclipse its bigger brother, so the 2.5-litre engine was detuned by lowering the compression ratio to produce a more modest output of 150bhp. Looking on the positive side, this was still 25bhp more than the 2-litre it was replacing, and performance was not significantly inferior to that of the wider-bodied 944.

The new engine transformed the 924. Suddenly it was powerful, smooth and, thanks to inheriting the 944's suspension developments, more confident in its handling. For the 1988 model year, towards the end of production, the differential in power was finally eliminated, making the 924S the better car, apart from its lack of the 944's more modern equipment.

That was the good part.

The 924S still demonstrated why customers were becoming frustrated by all of Porsche's four-cylinder cars. Now fully ten years into production, all their complaints should have been addressed, particularly as Porsche had put up prices sharply. For customers finding it hard to see what extra value they were getting for their money, it seemed that Porsche was beginning to slightly lose its way. The standard model did not even have a mirror on the passenger door or wiring for a stereo/cassette, while in right-hand drive markets customers still had to accept wipers that were set for left-hand drive and a bonnet release in the passenger footwell.

It was clear that Porsche was taking customer loyalty too much for granted, especially at a time when Mazda's RX7 was winning friends everywhere. Suddenly the writing was on the wall. In 1987 the previously steady demand dropped sharply and in the following year the 924S was discontinued.

It still remains, however, the best of the 924

The 924S, the pinnacle of 924 development, combined the classic good looks of the original design with the lively 2.5-litre all-aluminium engine of the 944 – the change in character was total. It was difficult to distinguish from the 2-litre model from behind – but its discreet potency could be a useful advantage.

924S

models with, not surprisingly, the 1988 160bhp version the one to have, particularly if it is the uprated Le Mans limited edition.

Bodyshell

European 924s had been given the Turbo's small rear lip spoiler, which was carried over to the 924S to reduce the drag coefficient from 0.36 to 0.33, as well as significantly improving the car's looks. During the 1982-84 period all Porsche's development effort had been concentrated on the 944, meaning the 1986 924S had no major bodyshell changes from the 1985 2-litre 924. Colours were also identical, both to the previous year and the 944, but there was the important change of the extension of the bodyshell warranty from seven to ten years across the whole range.

The 924S continued to use the 924's 66 litre (14.5 Imperial gallons, 17.4 US gallons) fuel tank, rather than the 84 litre (18.5 Imperial gallons, 22.2 US gallons) tank fitted to the Turbo and 944. But the tank was modified by adding an active carbon filter canister in the vent to absorb fumes.

Body Trim & Fittings

The only difference from the 1985 model in terms of trim was the new '924S' logo on the rear. And the only appearance upgrade in the 924S's life came in 1988 when, for the 160bhp version, an unattractive, flexible polyurethane lower edge was fitted to the front valence under the radiator opening. This could not really be called a spoiler or air dam as its lower edge did not project forward.

Interior Trim

The interior development of the 924 had pretty much stood still after the 944 had been announced, with few changes between 1982-84. However, 1985 saw the introduction of a new colour card and subtler tones to the interior, with the deletion of Pasha fabric. For 1986 the only further change was the deletion of the colour option of burgundy.

Although there was as much sound-deadening as in the outgoing 2-litre, this was still less than the 944, meaning the 924S was susceptible to noticeable road noise and irritating, even if unimportant, rattles from the transmission.

For taller drivers, access to the driver's seat was, if anything, slightly more difficult than before as the seats were set slightly higher. Using the low-mounted 380mm (15in) steering wheel supplied as standard, it was possible for hands to bump into legs when turning the wheel. As with the 924, the optional 360mm (14in) steering wheel was a good choice for taller drivers.

Tail of the 924S has two reminders, one less subtle than the other, of the manufacturer. The small badge is embossed in the lip of the spoiler.

Dashboard & Instruments

By 1986, the 944 had received the much improved 'oval' style dashboard. However, to differentiate the 924S downwards from its more expensive stable-mate, the original three-hole instrument layout was carried over from the 1985 924.

This was not a wise move. All the criticisms about difficulty in reading the instruments, especially with the smaller steering wheel fitted, continued, although changing the instrument markings to white from the previous yellow helped with the bits you could see. Meanwhile the 924S was also not given the mid-1985 upgrade to the 944's heating and ventilation system, meaning it inherited the problem of continuous heat output, even on the lowest setting.

From the first to the last, the most important asset of the 924 was its practicality. The rear seat back drops forward by releasing the two buttons on the top. Note the extractor vents for internal air circulation at the lower edge of the hatch sides.

51

Original Porsche 924/944/968

The 924S interior was carried over directly from the outgoing 2-litre 924, meaning the new oval dashboard of the 944 was not used. After ten years of improvement, however, the 924 cabin was a comfortable place and touring was a relaxing experience. Rear seat accommodation is excellent for children, and adequate for adults on short journeys. This 924S has lap straps only, but three-point belts can be fitted by appropriately qualified specialists. Load capacity increased with the seat back folded forward, and the area under it was useful for concealing valuables.

924S

The 924S adopted the attractive, smaller, four-spoke steering wheel first seen on the 924 Turbo, but you still had to look round it to see what was going on. At least the instrument faces were improved, with larger numerals.

Luggage Compartment

There were no changes from the 1985 2-litre 924, although in 1988 the 944's new split rear seat backrest and luggage cover became an option. This was not considered as good as the original, and very effective, roller blind cover.

Engine

A full description of the 2.5-litre water-cooled engine is given in the 944 chapter (see page 67).

For the 924S, several changes were made to reduce power output, as well as standardising lead-free fuel requirements and emissions equipment. The 150bhp engine was the same type as used in the US market between 1985-87. Compression ratio was 9.5:1, which allowed the car to run on 91 RON unleaded fuel.

for all world markets. The 150bhp 924S engine was the same as that used in the previous year's US 944 models, giving 147hp (SAE) net and designated M44/07, or M44/08 for the automatic transmission version. This was the first Porsche to have the same quoted power output for all markets, regardless of whether or not a catalytic converter was fitted. For the US market the only difference was a revised fuel injection programme for the engine management system to meet the Lambda control requirements.

To bring the output down to 150bhp, a new engine management chip was installed, along with a reduction in compression ratio from 10.6:1 to 9.7:1, achieved by deepening the bowl in the top of the pistons. The cylinder head remained unchanged from the one fitted to the 944.

Thereafter any annual changes to the 944's engine were automatically applied to the 924. For 1987, the 928-derived automatic tensioner was fitted to the timing belt at the front of the engine, eliminating the previous knuckle-scraping performance required for either adjustment or replacement. There was also an improved oil pressure relief valve and a common dipstick across all the four-cylinder models.

For the 1988 model year, the 924S and 944 engines became identical, the 924S benefiting from an increase of 10bhp, while the 944 dropped by 3bhp. The change was achieved by using identical pistons which gave a compression ratio of 10.2:1 and consequently raised the 924S's original fuel octane rating from 91 RON to 95 RON. This year also saw further standardisation across all the four-cylinder models with the use of the same connecting rods and exhaust systems, with or without a catalytic converter. Meanwhile a new sensor in the sump gave advance warning of low oil level, which was flagged alongside the other fault warnings on the instrument cluster.

Transmission

The 924S's driveline was identical to the 944's and used the five-speed Audi-derived gearbox, either 016J or 016K, according to market. The 1987 model year saw changes to the transaxle casing for the limited slip differential, with the rear axle strengthened by using new drive couplings manufactured by Oerlikon.

All markets had the same three-speed 087M automatic transmission unit.

Electrical Equipment

There were no specific changes to the 924S from the outgoing 924. The 944's standard 90 amp alternator (1260 watt) was continued, along with a standard 50Ah battery (63Ah with automatic transmission). US customers got the heavy-duty 115 amp alternator with a maximum output of 1610 watts, along with the larger 63Ah battery, in order to cope with the demands of a higher basic specification including air conditioning, power steering and electric windows.

Yet, even though Porsche was on a roll in 1986, the level of standard equipment was surprisingly low, with a passenger door mirror and central locking only offered as options – poor forward thinking which, when combined with poor value for money, was to bite the company badly in years to come.

Suspension & Steering

The front suspension was taken from the 944, with the bonus of the cast lower front wishbones which had been fitted in 1986. The rear suspension also adopted the 944's new cast semi-trailing arms. The first 924Ss had a 20mm (0.79in) front anti-roll bar as standard, with an optional sport specification of 21.5mm (0.85in) for the front bar and 14mm (0.55in) for the rear one.

The steering rack ratio was increased from the 924 Turbo's 20:1 to 22.39:1, mainly to lighten the effort when there was no power assistance. Power-assisted steering was optional and thought very desirable with the wider 195/65 tyres fitted as standard. Several markets, including the US, therefore adopted it as standard specification from the start or, in the case of the UK, for 1988.

Brakes

In recognition of the 924S's greater power, it was given the 944's brakes, originally derived from those of the 924 Carrera GT. The 9in (229mm) servo was used in conjunction with a tandem master cylinder operating floating calipers on ventilated discs for all four wheels. Front discs were of 292.1mm (11.5in) diameter and the rears 289.6mm (11.4in).

Unlike the 944 Turbo there was no rear circuit pressure relief valve to prevent rear wheel locking under heavy braking. And, sadly, the Anti-block Braking System (ABS) was never offered.

Wheels & Tyres

The 924S was given a new look compared to the 924 by having cast alloy 6J × 15 wheels with five oval holes. Road testers immediately dubbed them 'telephone dial' wheels, and the expression stuck. Tyres were 195/65VR15, a good size as they gave improved grip over the earlier 185s, but made for lighter steering than the optional 205/55s on 928-style forged alloy 6J × 16 wheels.

A collapsible Space Saver spare wheel was

924S

From the side the 924S was easily identified by its cast alloy wheels – the 'telephone dial' design is well-suited to the classic lines of the 924. This wheel has a locking wheel nut on the top stud.

progressively introduced to the 924 in many markets from 1980. It was included as standard from the 1988 model year for all UK cars.

924S Le Mans (1988)

To boost the 924S's image, a Le Mans special edition was built during the 1988 model year to celebrate Porsche's 12th outright win in the famous 24 Hour race. The factory order number for this model was M755 for all markets except the US, where it was M756.

The Le Mans was lowered 10mm (0.39in) at

The 160bhp 924S Le Mans celebrated Porsche's 12th outright win at the famous 24 Hours. It shared the same engine as the 944, but was lighter, offered less drag and its suspension was lower and stiffer. In short, it was the better car. These cars show the two colour finishes – Alpine White with yellow ochre detailing and Black with turquoise detailing.

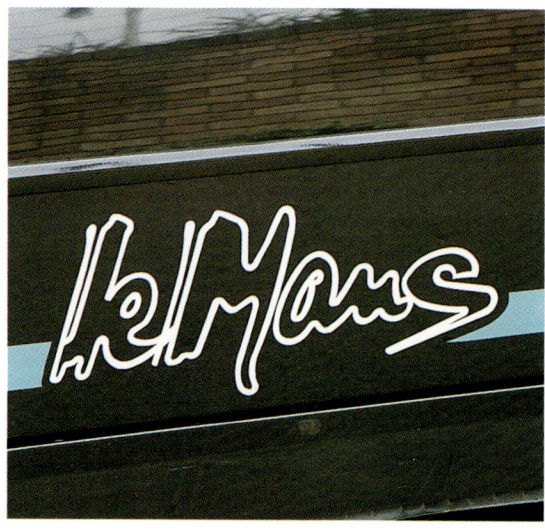

Original Porsche 924/944/968

Colour detailing on the 924S Le Mans models revealed in a cast alloy 'telephone dial' wheel and in the side logo that came with these cars unless a customer specified its deletion.

Fashionable yellows and greys set the interior of the Alpine White 924S Le Mans models apart from the crowd.

the front and 15mm (0.59mm) at the rear, and fitted with stiffer springs and gas-filled shock absorbers all round. It was further given sport anti-roll bars with diameters of 21.5mm (0.85in) at the front but 20mm (0.79in), rather than 14mm (0.55in), at the rear. Wheels were 'telephone dial' cast alloy 6J × 15 at the front and 7J × 15 (M396) at the rear. The 160bhp engine was unchanged.

The special paint finishes and interiors only offered two colour choices – Alpine White with Ochre/Grey detailing or Black with Turquoise detailing. The exterior side stripes were broken by scripted 'Le Mans' logos on the lower part of the door, while the rims of the holes in each wheel were in the appropriate Ochre or Turquoise.

Inside, the cars featured sports seats. Upholstery was grey/ochre striped flannel cloth with ochre piping for Alpine White cars, or grey/turquoise with turquoise piping for Black cars. All the cars had the 360mm (14in) steering wheel, with, as a final flourish, the cloth door panels colour-coded to match the detail colours. All cars came with an electric tilt/removable sunroof, normally an option, fitted as standard.

Between July and September 1987 a total of 980 924S Le Mans models were manufactured: 200 black and 50 white cars were sold in West Germany, 500 black (but no white) in the US, and 113 black and 117 white elsewhere in 'Rest of the World' markets.

56

Data Section

IDENTIFICATION

Chassis numbers For a full explanation see the 924 chapter (pages 31-32).

Engine numbers The first engine type in the 'Identification' table is for manual transmission models (eg, M44/07). The second is for automatic transmission (eg, M44/08). The 150bhp engine serial numbers started 43, followed by the model year letter (G for '86, H for '87) and then a five-digit build number (eg, 43G00001). The 160bhp engine serial numbers changed to 46 followed by the model year (J) and the five-digit number. Engines for the auto transmission have the five-digit numbers, starting with 6 (eg, 60001).

The 150bhp engine equates to 147hp net SAE. The 1988 160bhp engine equates to 154hp net SAE.

Gearboxes The optional limited slip differential on the 016J unit (Europe/RoW) is denoted in the transmission serial number by the first two digits (7Q), as opposed to the non-LSD gearbox QK (followed by the date of manufacture). A typical gearbox serial number might be QK19026. This means the unit is an 016J (Europe/RoW) with LSD, manufactured on 19 February 1986. From the 1987 model year US/Canada/Japan models used the same transmission, now called just 016.

The automatic version of the 924S used engine designator M44/08 with the 087M three-speed auto transmission with serial numbers starting RCD (Europe/RoW), followed by the date of manufacture. For 1987 the automatic transmissions were different, the US/Canada/Japan version being the 087N, with serial numbers starting RCB.

One of the most desirable of the 924 models – the 924S Le Mans limited edition from the 1988 model year. Only 980 were built.

IDENTIFICATION DATA

Year	Model	Market	Chassis numbers	Engine type	Gearbox type
1986	924S	Eur/RoW	WPOZZZ92GN400001-3536	M44/07,08	016J, 087M
1987	924S	Eur/RoW	WPOZZZ92ZHN400001-1993	M44/07,08	016J, 087M
	924S	US	WPOAA092_HN450001-6947	M44/07,08	016K, 087N
1988	924S	Eur/RoW	WPOZZZ92ZJN400001-2003	M44/09,10	016K, 087M
	924S	US	WPOAA092_JN450001-2190	M44/09,10	016K, 087N

PRODUCTION DATA

Year	Model/market	Max power (bhp@rpm)	Max torque (Nm@rpm)	Compression ratio	Weight (kg)	Number built
1986	924S Eur/RoW	150@5800	195@3000	9.7:1	1190	3536
1987	924S Eur/RoW	150@5800	195@3000	9.7:1	1190	1993
	924S US	150@5800	190@3000	9.7:1	1240	6947
1988	924S Eur/RoW	160@5900	214@4500	10.2:1	1240	2003
	924S US	160@5900	214@4500	10.2:1	1240	2190
Total						16,669

KEY FACTS

Fuel 91 RON (1986-87), 95 RON (1988).

Dimensions Length, 4212mm (US, 4290mm); width, 1685mm; height, 1275mm; ground clearance, 120mm; turning circle, 10.3m; weight (kerb), 1190kg (US, 1240kg).

Capacities Engine oil, 6.0 litres; gearbox, 2.0 litres (manual) or 6.0 litres (automatic); fuel tank, 66 litres; screenwash, 6.0 litres.

OPTIONS

See 944 list (pages 73-74).

COLOURS & INTERIORS

1985
See 924 section (page 33).

1986-88
See 944 section (page 74).

Porsche 924/944/968

The 944 (1982-89)

In 1980, Porsche's fortunes were at a low ebb. Production of the 924 was slowing and even the announcement of the Carrera GT and the entry of three prototypes for the Le Mans 24 Hours had failed to revive flagging interest.

The company's strategy was now framed entirely around the new water-cooled cars, led by the 928, with plans to phase out the 911 by about 1984. In what was possibly a calculated move to stimulate continued interest, news of a new engine to be installed in the existing 924 had begun to leak from Weissach during the late '70s. Observers have commented that the object of this policy was to take a larger share of an existing market. Yet, as that itself was contracting, it was a poor strategy.

The company needed a new broom, with new

Evolution Outline

Jun 1981	163bhp 944 announced.
Nov 1981	Production starts.
Apr 1982	UK launch.
May 1982	US launch.
Aug 1983	Electric release for rear window, optional electric tilt sunroof.
Aug 1984	Power steering becomes standard, 'telephone dial' wheels introduced.
Feb 1985	Major upgrade, including new oval dashboard, optional central locking.
Aug 1987	Same 160bhp model for all markets.
Aug 1988	Engine capacity increased to 2.7 litres.
Jul 1989	Discontinued.

944

ideas, and the man chosen was the German-American Peter Schutz. As soon as he took office on 1 January 1981 he saw the potential for the new engine, then being developed in a bodyshell derived from the 924 Turbo. But Schutz was looking for expansion rather than consolidation, and saw giving the new model a bodyshell derived from the Carrera GT as an expedient way of attracting new business.

With a new model number of 944, the car was an immediate success. Even if its parent was easily identifiable, the launch was pitched at presenting the 944 as a brand new model to try to finally bury the image of the car not being a 'real' Porsche. This policy worked better than Schutz could ever have imagined. While Porsche, as a company, built a *total* of some 28,000 cars in 1981, just two years later it had built over 26,500 944s alone, 16,618 of them for the US.

Schutz was not aiming for Porsche's traditional enthusiast buyers with the 944. Instead he was after a new cash-rich type of customer, attracted

From February 1985, the 944 was significantly upgraded. Externally there was improved anti-corrosion treatment to the underside and a flush-mounted windscreen. This car has the rather expensive – and very desirable – option of forged alloy Fuchs wheels.

Porsche 924/944/968

by the car's aggressive good looks. Major market research was carried out into the car's suitability for the US market, with the projected figures then receiving a huge boost from the favourable exchange rate between Germany and the US.

The 944 became one of Porsche's most significant models ever, breaking all sales records and generating substantial cash reserves. These would buoy the company during the dark years of the late '80s and early '90s, along with a further cash injection from the 1984 public flotation of stock, although this was still non-voting – power remained with the Porsche and Piëch families.

These new resources led to significant investment in modernisation and new production facilities, yet, unfortunately, at the same time there was complacency about product development. As a result sales plummeted when the world recession hit after 1987, especially as exchange rates also turned round. It was not to be until 1994 that the company managed to produce, in the 993, a car with the potential to reverse the decline. In the meantime it could thank sales of the 944 for giving it the flexibility to survive.

The 944 represents incredible value today. It was significantly upgraded during 1985, benefiting from many of the improvements introduced for the new 944 Turbo, while the last 1989 models – particularly the 2.7-litre, which was not available in the US – represent the best of the range by virtue of their improved mid-range flexibility.

Bodyshell

The 944 bodyshell evolved from the 924 Turbo, except for the obvious external differences derived directly from the Carrera GT and including the distinctive flared front and rear wings. The deeper front valence, made of glass-fibre, featured two long rectangular openings for the entry of cooling air, rather than the 924 Turbo's single large opening, along with multiple grilles for additional radiator and brake cooling. The bumpers were tidied up in the wind tunnel so that the front valence integrated better at the lower edge, while the seam with the front wing was moved round from the more corrosion-exposed front face to the side location of the rubbing strip. An optional variant was a front valence with provision for two driving lamps, which were set into the moulding just below the elastomer front over-riders.

The result of all this was a significantly cleaner and more aerodynamically efficient profile to the front of the car. Considering the wider body, the

The inspiration for the 944 was the 924 Carrera GT. Wider flares on the wheel arches pushed the drag coefficient up from 0.33 to 0.35, but this was a small sacrifice for the improved looks.

944

Porsche 924/944/968

The 2.7-litre 944 was noticeable for its improved torque and flexibility, rather than extra aggressiveness – but there was no change externally. The 944 kept the separate bumper assembly all through its production life, making it easy to distinguish it from the 944S2 and Turbo models.

The squared-off wing flares are typical of the early '80s, but suit the 944 perfectly. At the rear they have the added advantage of allowing wider rear tyres to be accommodated.

The small side repeaters behind the front wheels were fitted only to the non-US specification models. Note the colour-coded rubbing strip.

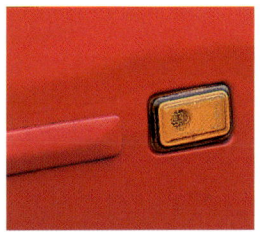

The bonnet finally received gas strut supports on the 944, replacing the cheap rod support of the earlier 924s. These early struts become tired over the years, but they seemed like progress at the time!

drag coefficient of 0.35 was impressive, slightly better than the original 924's 0.36 although still short of the 924 Turbo's 0.33.

Porsche chose not to use the 924 Turbo's NACA bonnet intake, although the 1981 Le Mans 924 GTP, the 924 most closely resembling the 944's new shape, did have it. For all 944s, including the Turbo, the bonnet remained clean.

The clean lines were carried over to the new rear wings, integrated into the existing adjacent sections of bodyshell. As with the 924, the front wings were removable, for easy accident repair, with both front and rear units being made from zinc-coated steel. The zinc coating process was so effective that from 1986 the anti-corrosion warranty was increased from seven years to ten.

The lower front cross-member, made from light alloy, carried the inner ends of the lower front wishbones, the engine mounts and the flexibly mounted steering rack, the latter contributing towards reducing acoustic noise. The considerable attention paid to reducing interior noise also resulted in new engine mountings which used glycol-filled units acting like shock absorbers, one each side of the block and mounted onto the cross-member. However the unit near the exhaust was soon to give problems and eventually all 944s used a much more robust engine mounting developed for the later Turbo. The gearbox and steering rack were also elastomer-mounted to reduce noise still further.

For the mid-1985 upgrade, the 944 duplicated the Turbo's three-point mounting of the engine, torque tube and gearbox, further reducing interior noise, while a new 80-litre (17.6 Imperial gallons, 21.1 US gallons) polyethylene fuel tank gave significant extra capacity over the old 66-litre (14.5 Imperial gallons, 17.4 US gallons) steel tank.

No discussion of the 944's body shape can be complete without mentioning Guards Red, a colour which became synonymous with the model in representing the loud and flashy look of the '80s. Now of course rather dated, it is the colour which best gives the car the right 'period look'.

Body Trim & Fittings

The add-ons given to the 924 over its relatively short life – at least by Porsche standards – were tidied up on the new 944. The ugly front and rear

Porsche 924/944/968

Only one digit changed for the badge on the tail, but the 944 represented a huge step forward for the water-cooled four-cylinder Porsches. All sorts of details typified the new model's improved quality and character.

A brand new front spoiler and a trimmed front bumper ensured the 944 looked neater from the front, and lighting improvements were long overdue. On this European specification car fog lamps in the lower valence augment the driving and main lamps.

The rear spoiler (below) was larger on the 944, matching that first seen on the Carrera GT. The electrically tilting sunroof (bottom) was an option from the 1984 model year. The roof could still be removed completely and stowed in the luggage compartment.

side repeaters of the US models were replaced with narrower and more stylish units at bumper level. The bonnet received new gas struts to replace the previous folding metal prop, while the electric driver's door mirror was made a standard factory fitting. For 1985, the bonnet was strengthened and the strut arrangement further improved.

Underneath, Würth PVC underseal was applied before the final body coat, appearing under sill level between the wheels as a stippled finish. The 924's narrow black rubbing strips were replaced by colour-coded strips at waist level between the wheels. Apart from their obvious function, they were important stylistically in breaking up the large expanse of bodywork below the side glass. The rear spoiler, still manufactured from flexible black elastomer, was enlarged, while a '944' logo replaced the earlier model designation, still using the block style fashionable on '70s and early '80s Porsches.

For the 1985 upgrade one of the most visible improvements was flush mounting the front windscreen with the bodyshell to replace the previous recessed arrangement. Separate options were a Sekuriflex laminated windscreen and one with a graduated tint. On right-hand drive models the wipers at last parked on the left. From the start of the 1988 model year the wipers improved still further by sprouting small spoilers to make them more efficient at speed.

From the start of the 1986 model year central locking became an option across the range.

944

Interior Trim

The high-backed seats offered across the entire model range were also standard on the 944. As the trim developed these seats became available in increasingly attractive fabrics (alongside leathers to special order), with the most popular interior choices either the new Berber cloth or the pinstripe. Carpets also became plusher.

The first 944s carried over the improved dashboard layout of the 1982 924, and used the four-spoke steering wheel first seen on the 924 Turbo. Improved numbering was used on the main instruments, and all markings were now in yellow.

For the 1983 model year speaker covers were fitted to the doors, while 1984 introduced the tasteful 'Porsche' lettered cloth in black, brown or grey-beige. In conjunction with the 1985 dashboard upgrade (see below), the 360mm (14in) diameter steering wheel was raised by 18mm (0.71in) and the seat was lowered by 30mm (1.18in), adding much-needed comfort for tall drivers as well as providing significant extra room for getting in and out. From 1988, a tape and coin holder was fitted to the rear of the central tunnel.

Dashboard & Instruments

The original dash was identical to the improved one fitted to the 924 for 1982. It featured updated instrument graphics and a 10% increase in ventilation air throughput by using a radial fan and revised ducting. A fuel consumption meter, located in the tachometer, was an option.

The major change to the interior came in February 1985, with a complete revision of the dashboard and centre console. The upgrade introduced the dash assembly, controls and instruments first seen on the new 944 Turbo. The instruments themselves now had white or red markings on

Porsche 924/944/968

The February 1985 upgrade introduced a completely new – and much improved – dashboard and centre console layout. There was extra leg-room under the steering wheel, but the view of the neat new instruments was still not wonderful. The February 1985 upgrade also introduced new door trim, tidying up the window and exterior mirror controls.

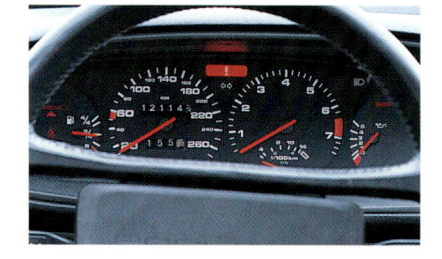

These views show later 944s with optional air conditioning (left) and with the standard fully automatic climate control (above). The latter provided a 'dial-a-temperature' facility which worked very well, but interior temperature, of course, could only be controlled down to the outside ambient. Other new features were the digital clock (which optionally showed outside air temperature) and the internal air recirculation control. Maximum air flow for ventilation increased by 35 per cent.

black backgrounds and, for automatic models, the shift position was now indicated on the tachometer. Larger fresh air vents contributed to a claimed 35% increase in air flow, while heating and air conditioning – where that was fitted – were also revised, improving performance by a claimed 10% and 17% respectively.

As well as giving a very 928-like appearance, the new oval-shaped dash at last addressed the problem faced by taller drivers by being introduced in conjunction with the changes to steering wheel and seat positions mentioned in the 'Interior Trim' section (page 65).

Luggage Compartment

At launch, the luggage compartment was identical to that of the 1982 924 Turbo and was given few later improvements, apart from trim upgrades. However, the 1989 model year saw the dubious change to split rear seat backs.

This arrangement was useful in that it allowed long items to be carried while keeping one rear seat available, but it also meant the replacement of the excellent roller blind over the luggage area by a rather inadequate curtain-like affair which hooked to the rear glass.

Engine

There were plenty of rumours in 1979 that a new engine for the 924 was in development, based on the knowledge that Weissach wanted to use the technology developed on the 928. Inevitably robust discussions must have taken place as to whether this should be an in-line four, a V6 or a straight-six. But in the end practicalities drove the decision. A straight-six would not fit into the 924's engine compartment, while a V6 could not use the identical cylinder head from the 928 – an important factor for a relatively low-volume producer like Porsche. In the end, therefore, the 'four' was chosen, giving the engineers the opportunity to show they could bring this layout to the same pitch of refinement and smoothness as a 'six'.

The new all-aluminium engine had a capacity of 2479cc (151.27cu in) and was inclined at 30° – rather than the 40° of the taller 2-litre – to fit the engine bay. Bore and stroke were an oversquare 100mm (3.94in) by 78.9mm (3.11in). The stroke was identical to the 928, as was the bore spacing, but the actual bore would only be copied on the 928 for the larger 1986 S4 model.

Given the continued criticism of the roughness of the 924's 2-litre engine, special attention

The 944's new 2.5-litre all-aluminium engine was tilted at 30° to fit the existing space. Damping of the engine's vibrations to the bodyshell was reduced by fluid-filled engine mountings.

Porsche 924/944/968

was paid to engine balance. The balancer shaft principle had first been proposed early in the evolution of the motor car by Lanchester, but after considerable study Porsche chose a two-bearing derivative developed later by Mitsubishi in Japan. The Lanchester principle overcomes unbalanced forces through two counter-rotating parallel shafts running at twice the crankshaft speed. The shafts balance not only the major oscillating and rotating components, so as to eliminate the primary free forces, but also secondary inertia forces caused by vibration from the up and down movement of the pistons and connecting rods. These shafts, one each side of the block and driven by belts from the crankshaft, have calculated imbalance, thereby giving significant overall smoothness throughout the rev range. The price, however, is 5bhp being needed to drive the system.

The forged crankshaft ran in five main bearings, with oil circulated by a crescent-type pump. The crankshaft was given great stiffness by being located by a large aluminium ladder frame bolted to the underside of the main block. As with the 924, the sump pan was a heavily-baffled aluminium casting, which helped overcome the effects of oil surge. A novel oil/water intercooler was located on the exhaust side of the block, an arrangement which initially proved fragile. This helped the oil to warm up on cold days and cooled it on hot days or at high engine speeds. The aluminium-silicon alloy block was linerless, with the pistons iron-sprayed to prevent cohesive pick-up in the bores.

The new cylinder head was a combination of lessons learned during the early development of the aluminium V8 for the 928 and the Thermally Optimised Porsche (TOP) cylinder head study of 1980-1, which considered the extent to which compression ratio, combustion chamber design, air/fuel mixture and ignition timing could be optimised together to give better performance, fuel consumption and exhaust emissions. The head was designed for significant commonality with the 928, although at first the shape of the combustion chamber was slightly different. A single belt-driven overhead camshaft operated two in-line valves per cylinder through hydraulic tappets. Compression ratio was 10.6:1 for 98 RON petrol in Europe or 9.5:1 for 91 RON lead-free in the US.

The Digital Motor Electronics (later to evolve into Motronic) integrated engine control was fitted, along with the improved Bosch L-Jetronic fuel/air delivery system. US versions of the engine had Lambda control.

Although the new engine was slightly heavier than the old 2-litre, with a dry weight of 154kg (339lb) compared to 136kg (300lb), it offered much smoother power delivery, greater fuel efficiency and cleaner exhaust emissions. Output was 163bhp (DIN) at 5800rpm for Europe, and 143hp

The cylinder head (top), designed for a high degree of commonality with the 928, adopted lessons learned with the Thermally Optimised Porsche (TOP) design study of 1980. An aluminium cover (above) kept most road debris out of the engine bay. Working from underneath was the best way into the camshaft belt tensioner of the early cars.

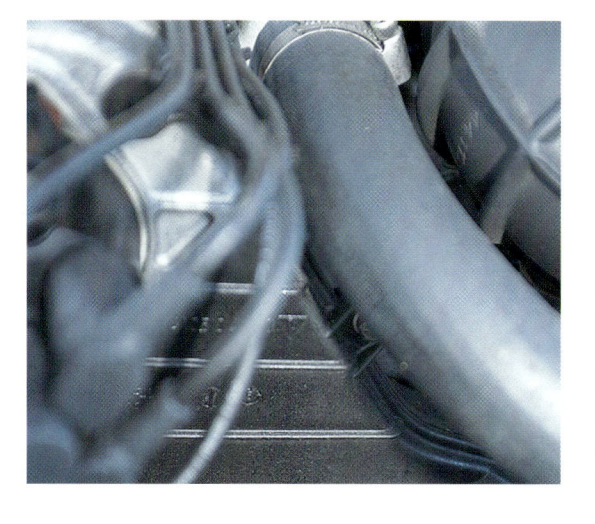

A toothed belt drove the single overhead camshaft. Under the plastic covers are also the two balance shaft drive belts – the secret of the 944 engine's smoothness.

944

The 2.7-litre engine was typically under-stated. Although maximum power and torque increased by only 5bhp and 11lb ft, it was a much more flexible unit in the lower revolution ranges. The cylinder block technology was derived from the TAG F1 programme.

(SAE) at 5500rpm for the US. Torque was especially impressive. Whereas the 924 had peaked at 165Nm (122lb ft) at 3500rpm, or 151Nm (111lb ft) for the US model, the new engine produced more sustained torque delivery which peaked at 205Nm (151lb ft) at 3000rpm, or 186Nm (137lb ft) for the US. The engine had 12,000-mile service intervals, like the 911 and 928.

Over the first few years of production the engine was then further developed in detail. The cam and balance shaft drives were improved, particularly to extend the unacceptably short lives camshaft belts were proving to have. For 1984, there was a reworking of oil circulation in the crankcase and its ventilation, with larger borings lubricating the main bearings. A pressure attenuator was installed in the fuel collection pipe to reduce injection pulse vibrations.

The 1985 upgrade included revised connecting rod nuts, an uprated oil pump giving a 10% greater flow rate through larger oil passages, and an oil capacity increase from 5.5 litres to 6 litres. Revisions to the combustion chamber shape included new exhaust valves, and new pistons were fitted. The DME electronics were upgraded and there was a revised air mass measurement system. There was also a larger water radiator.

For 1986, the exhaust valves were made from harder material and the exhaust manifold manufactured from sheet steel. An active carbon filter was also mounted in the fuel filler of models fitted with a catalytic converter. In 1987 more changes followed to the drive belt arrangement, including an automatic camshaft tensioner, along with an improved oil pressure relief valve.

In 1988 the differences in output between the various markets were eliminated by standardising it at 160bhp, with or without catalytic converter (respectively the M44/09 and M44/10 engines). Compression ratio was reduced to 10.2:1 so that all engines could run on 95 RON lead-free fuel. A new sender in the sump advised of low oil level.

Porsche 924/944/968

At the end of the 1988 model year the 944 was discontinued in the US, but for 1989 – the final year of the regular 944 in other markets – capacity was increased to 2681cc (163.6cu in) by enlarging the bore size to 104mm (4.09in), using the new cylinder block described in the 944S chapter (see page 93). Compression ratio was raised to 10.9:1, the diameter of the inlet valves increased from 45mm (1.77in) to 48mm (1.89in), the inlet ports enlarged, the camshaft timing revised and the water pump flow rate increased.

The engine management system was also changed to enable all models to continue running on lead-free petrol. By now they were all also fitted with a three-way catalytic converter, Lambda probe and active carbon filter to absorb fuel fumes. There was also a new version of the DME, with fuel system pressure increased from 2.5 bar to 3.8 bar through a new fuel pump, alongside revisions to the pressure regulator, fuel attenuator and injection valves.

The overall increase in power, to 165bhp, was disappointing, but there was a noticeable improvement in torque, now raised to a maximum of 225Nm (166lb ft) at 4200rpm.

Transmission

Power transmission from engine to gearbox was identical to the 924 Turbo. A single 225mm (8.86in) diameter hydraulically operated dry-plate clutch transmitted torque through a 25mm (0.98in) diameter drive shaft, supported in three sealed-for-life bearings.

At first there were two variants of the 016 five-speed manual gearbox, produced, as before, by VW-Audi in Kassel – the 016K for the US and Japan or the 016J for other markets. Fourth and fifth gears were longer than the 924 while, at 3.889:1 (9/35), final drive was the same as the Series 2 924 Turbo but longer than the regular 924. A limited slip differential was optional. As with the 924 Turbo, double universally jointed drive shafts transmitted power to the wheels.

The transmission casing was revised for the 1985 model year, with provision for mounting an electronic speedometer to sense gear speed. For the 1988 model year, in line with standardising the engine, the 016J, coupled to a 3.889:1 final drive, was made the only manual transmission.

The three-speed automatic – 087N for the US and Japan, 087M elsewhere – was unchanged from the 924, but final drive was lengthened to 3.455:1 (11/38) for the 087N or 3.083:1 (12/37) for the 087M. For the 1983 model year (actually from 1 October), first and intermediate were shortened on 087M, improving acceleration from a standing start. The gearbox serial number prefix also changed, from RH to RCA, followed by the date of manufacture. The two market variations were maintained to the end of 2.7-litre production.

There were several subsequent detail changes for the automatics, such as the 'in-gear' indicator in the tachometer and an improved shift lever for 1985. On the manuals, there were revisions for 1987 to the limited slip differential, along with improved drive shafts.

Electrical Equipment

The 12-volt system was virtually identical to the 924, with European and Rest of the World models using a 50Ah battery, turned through 90°, and US models having the optional 63Ah as standard. Alternator output was 90 amps and 1260 watts. By 1989, for the more complex 2.7-litre model, the battery was standardised at 63Ah, while alternator capacity rose to 115 amps and 1610 watts.

For 1983 there was an improvement across the range in the radio equipment specified by the factory. US models were fitted with the Blaupunkt Monterey SQR22, other markets with the Cologne version. As with the 924, the sunroof could be tilted electrically and the tailgate released from the driver's seat. A new option for 1984 was cruise control, while the electric windows could now be operated with the ignition key removed and the doors open.

For 1985, a lighter and less noisy starter motor was installed. There were new electrically heated windscreen washer nozzles and, for the first time, a radio aerial integrated into the windscreen. Other interesting new options included central locking, heated front seats and electric height adjustment for the front seats. In the mid-year major upgrade, the fuse panel and relays were moved from the footwell to the rear of the engine compartment for better cooling, partly because the system was becoming exceedingly complex.

From 1988 central locking became standard for all markets, rather than for specific countries, along with electric windows, a passenger door mirror and four loudspeakers. Each market had its own policy when it came to radios, although the factory always offered the latest Blaupunkt. In the 1985 model year, for instance, the factory radio/cassette option was the Hamburg SQM24, with ARI (automatic traffic information). This was followed for 1988 by the Ludwigsburg SQM26 and in 1989 by the Bremen SQR46, both also with ARI. Even higher specification was optional through the Berlin digital self-seek unit and, separately, a ten-speaker system.

From 1988, German models could also be supplied from the factory wired for the national C-net cellular telephone system. Other markets offered similar arrangements for specific national cellular systems.

Suspension & Steering

The 944 had a standard 20mm (0.79in) front anti-roll bar, with some markets in addition offering a 14mm (0.55in) rear anti-roll bar as standard, although this was not an across-the-board fitting until as late as 1988. Cast alloy lower front wishbones and rear semi-trailing arms came with the 'oval dash' cars in 1985, stiffening up the dynamics of the suspension and also allowing easier adjustment at the front.

An option from the 1983 model year was variable assistance power steering, derived from the system used on the 928. This was first offered in conjunction with the option of automatic transmission, but soon became very popular on the manuals thanks to its progressive action, especially as it came with a steering ratio reduced from 22:1 to 18.5:1 to provide light steering which responded more quickly.

Many markets soon began specifying this power steering as standard in conjunction with the 363mm (14.3in) steering wheel which also improved leg room. After August 1983 power steering became standard on automatics and, after October 1984, on manuals too. A cooling pipe was added to the pressure circuit in 1985 to rectify problems with overheating pumps.

Brakes

The brakes were taken from the Carrera GT and used a 9in (228mm) diameter servo and tandem master cylinder. The discs, ventilated all round, were 282mm (11.1in) diameter at the front and 289mm (11.4in) diameter at the rear, the same as the 924 Turbo. From the 1984 model year, wear on the brake pads was monitored on all four wheels, a warning light indicating when they needed to be replaced.

For the 1987 model year, the Anti-block Braking System (ABS) was offered as an option. Fitting this involved complete revision of the braking system in order to accommodate the three-channel ABS installation and the sensors in each wheel hub. Revisions to the hubs included a wider front track, strengthened stub axles, larger wheel bearings, longer track rods, a dedicated front strut top support bearing, increased rear track and lengthened drive shafts. The ABS control unit itself was located in the right-hand footwell, with the hydraulic unit in the right-hand front wing.

For the 1988 model year the system was further improved and then, for 1989, made standard across the range. At the same time brake pedal effort on all four-cylinder cars was reduced through fitting a 3.4:1 ratio brake servo to replace the former 3.0:1.

MacPherson strut front suspension on the 944 (top) was derived directly from the 924 Turbo. Early 944s used the regular 924's fragile pressed steel fabricated lower A-arms (middle), which were susceptible to distortion against kerbs. The Turbo's cast lower wishbones replaced them in February 1985, improving handling stiffness. Rear suspension (bottom) was also taken from the 924 Turbo. After February 1985, the semi-trailing arms were stiffer aluminium castings.

Porsche 924/944/968

The front discs were improved for 1988 with revised ventilation passages, while asbestos-free pads were fitted for the legislative requirements of Sweden, Norway and Denmark. For 1989 these became standard across the range.

Wheels & Tyres

Launch models of the 944 were shown with traditional Porsche Fuchs 7J × 15 alloy wheels fitted with 185/70VR15 tyres. In addition there were two wheel options – 15in forged alloy rims all round with 215/60VR15 tyres or 16in versions with 205/55VR16 tyres.

For the 1984 model year, customers could have the centres of these optional wheels painted in special Grand Prix White or Pewter metallic, with Gold metallic replacing Pewter for 1985, when a third wheel choice was offered in the form of 7J × 16 with 205/55VR16 tyres for the front and 8J × 16 with 225/50VR16 tyres for the rear. Lockable wheel nuts were a further option.

The 7J × 15 cast alloy 'telephone dial' wheels, with 195/65VR15 tyres as standard, were introduced as part of the mid-1985 model upgrade which introduced the oval dash. From the start of the 1987 model year, fresh options were the 'telephone dials' in 7J × 16 and 8J × 16, or forged disc-style wheels. For 1988 customers could specify a platinum finish, rather than the standard silver, for the latter.

944 Celebration (1988)

By 1988 sales were falling dramatically, with production down by nearly a third from 1987. However, this only encouraged Porsche to celebrate the 944 breaking all the company's previous sales records. To mark the 100,000th car rolling off the Neckarsulm production line, a limited edition was made available in either Zermatt Silver or Satin Black metallic.

These 'Celebration' models – effectively standard cars brought up to a very high specification – featured black leatherette, the attractive grey/maroon 'Studio' cloth and silver grey carpeting. The new split rear seat backs were included, along with automatic heating control and removable electrically tilting sunroof. Wheels were the optional 7J front and 8J rear 'telephone dials', fitted with 205/55VR16 and 225/50VR16 tyres respectively. Integral fog lamps were also fitted at the front. US models did not have the rear '944' logo found on cars for other markets.

Standard 7J wheels changed in February 1985 from 'cookie-cutter' (top) to 'telephone dial' (middle). Fuchs forged alloy wheels (bottom) were a desirable but expensive option.

Data Section

IDENTIFICATION

Chassis numbers For a full explanation see the 924 chapter (pages 31-32).

Engine numbers Engines for 1982 with manual transmission were designated M44/0, with serial numbers starting 41 followed by the year letter (C for '82, D for '83, E for '84, etc), followed by the serial number (eg, 41C0001). US/Japan engines were M44/04 with serial numbers starting 43 followed by the year letter and serial number. The automatic engine was M44/03 (RoW) or M44/04. These were identified with a 5 after the year letter in the serial number, changing to 6 for 1985.

For 1985, the designators changed to M44/05 (RoW) or M44/07 (US) for manuals, 06 and 08 respectively for automatics. For 1988, engines were the same output for RoW and the US (M44/09 manual, M44/10 automatic) and were designated 46, followed by the year number (J for '88), followed by the serial number (with or without 6 at start of serial number). No 2.7-litre 944s were delivered to the US; 2.7-litre automatic transmission engines were M44/12.

Gearbox numbers From 1982 the 944's automatic gearboxes were as follows: 087M, coded RH (Europe/RoW until September 1982); 087M, coded RCA (Europe/RoW from October 1982); 087N, coded RCB (US/Japan from February 1982). The code numbers were followed by the date of manufacture in 'ddmmyy' format.

Manual transmissions were: 016J, coded QK (Europe/RoW); 016J, coded 7Q (Europe/RoW with LSD); 016K, coded QM (US); 016K, coded 8Q (US with LSD); 016K, coded QL (Japan); 016K, coded 4M (Japan with LSD).

KEY FACTS

Fuel European version: 98 RON. US version: 91 RON lead-free. From 1988, all models, 95 RON lead-free.
Dimensions Length, 4200mm (US, 4290mm); width, 1735mm; height, 1275mm; front track, 1477mm; rear track, 1451mm; wheelbase, 2400mm; ground clearance (empty), 138mm; turning circle, 10.3m; kerb weight, 1180kg (1982 Europe/RoW), 1267kg (1983 US auto), 1260kg (1988 Europe/RoW).
Capacities Engine oil, 5.5 litres (1982) or 6 litres (1985); transmission, 2.6 litres (manual) or 6.0 litres (automatic); fuel tank, 66 litres; screen/headlamp wash, 6.5 litres.

OPTIONS

Since 1982, option numbers have been included on the Vehicle Identification Number (VIN) plate. Some of these will be included in the factory basic specification. Note that full specification varied from market to market, as the basic factory specification was often supplemented by certain options which then became 'standard' in that particular market. With acknowledgement to Porsche Cars North America.

M18 Sport steering wheel with elevated hub
M20 Speedometer with two scales – kph/mph
M24 Version for Greece
M26 Activated charcoal canister
M27 Version for California
M30 Sport group with Koni shock absorbers (944)
M31 Sport group with Fichtel & Sachs shock absorbers (944)
M34 Version for Italy
M36 Bumpers with impact absorbers (924, 944)
M61 Version for Great Britain
M62 Version for Sweden
M111 Version for Austria (924, 944)
M113 Version for Canada
M119 Version for Spain
M124 Version for France
M126 Digital radio, 1982
M126 Stickers in French (924, 944)
M127 Stickers in Swedish
M130 Labelling in English
M139 Seat heating – left (924, 944)
M150 Without catalytic converter (924, 944)
M153 Engine parts specific to 944 Turbo
M154 Control Unit for improved emissions (924, 944)
M164 Tyres: 215/60VR15 (944)
M167 Bridgestone tyres
M182 Safety devices in passenger compartment (944)
M185 Automatic two-point rear seat belts
M186 Manual two-point rear seat belts
M187 Asymmetric headlamps
M190 Increased side door strength
M193 Version for Japan
M195 Prepared for cellular telephone system
M197 Heavy duty alternator 1988 (944)
M197 High amperage battery
M198 Stronger starter (1.7kW) for Canada (944)
M215 Version for Saudi Arabia
M218 Licence plate brackets, front and rear (US)
M220 Locking differential
M225 Version for Belgium
M240 Version for countries with inferior fuel
M249 Automatic transmission
M255 Fuel consumption indicator (944)
M258 Heating for outside mirror (924, 944)
M261 Passenger side mirror, electric, plain (924, 944)
M277 Version for Switzerland
M288 Headlamp washer

IDENTIFICATION DATA

Year	Model	Market	Chassis numbers	Engine type	Gearbox type
1982	944	RoW	WPOZZZ94ZCN400001-3921	M44/01,03	OK, RCA
	944	US/Can/Jap	WPOAA094_CN450001-	M44/02,04	QM, RCB
1983	944	RoW	WPOZZZ94ZDN400001-9127	M44/01,03	OK, RCA
	944	US/Can/Jap	WPOAA094_DN450001-65506	M44/02,04	QM, RCB
1984	944	RoW	WPOZZZ94ZEN400001-9921	M44/01,03	OK, RCA
	944	US/Can/Jap	WPOAA094_EN450001-66618	M44/02,04	QM, RCB
1985	944 Oval	RoW	WPOZZZ94ZFN400001-4967	M44/05	016J
	944	RoW	WPOZZZ94ZFN420001-3524	M44/06	RCE
	944 Oval	US/Can/Jap	WPOAA094_FN450001-9062	M44/07	016K
	944	US	WPOAA094_FN470001-6167	M44/08	RCF
1986	944	RoW	WPOZZZ94ZGN400001-6109	M44/05	016J
	944	US/Can/Jap	WPOAA094_GN450001-60901	M44/07	016K
1987	944	RoW	WPOZZZ94ZHN420001-2343	M44/05	016J
	944	US/Can/Jap	WPOAA094_HN470001-8246	M44/07	016Ks
1988	944	RoW	WPOZZZ94ZJN420001-2226	M44/09	016J
	944	US/Can/Jap	WPOAA094_JN470001-3731	M44/09	016J
1989	944	RoW	WPOZZZ94ZKN400001-2730	M44/13	016J

Note
The major specification upgrade, including the 'oval' shaped dash, came in February 1985. The production number split for 1985 is shown.

PRODUCTION DATA

Year	Model/market	Max power (bhp@rpm)	Max torque (Nm@rpm)	Compression ratio	Weight (kg)	Number built
1982	944 RoW	163@5800	205@3000	10.6:1	1180	3921
1983	944 RoW	163@5800	205@3000	10.6:1	1180	9127
	944 US/Can/Jap	143@5500[1]	186@3000	9.5:1	1267	15506
1984	944 RoW	163@5800	205@3000	10.6:1	1180	9921
	944 US/Can/Jap	143@5500[1]	186@3000	9.5:1	1267	16618
1985	944 RoW	163@5800	205@3000	10.6:1	1180	8491
	944 US/Can/Jap	143@5500[1]	186@3000	9.5:1	1267	15229
1986	944 RoW	163@5800	205@3000	10.6:1	1180	6109
	944 US/Can/Jap	143@5500[1]	186@3000	9.5:1	1267	10901
1987	944 RoW	163@5800	205@3000	10.6:1	1180	2343
	944 US/Can/Jap	143@5500[1]	186@3000	9.5:1	1267	8246
1988	944 RoW	160@5900	210@4500	10.2:1	1260	2226
	944 US/Can/Jap	160@5900	210@4500	10.2:1	1260	3731
1989	944 RoW	165@5800	225@4200	10.9:1	1290	2730
	944 US/Can/Jap	165@5800	225@4200	10.9:1	1290	2691
Total						117,790

Note
[1] Denotes hp SAE net, otherwise all figures bhp DIN. All weights are figures with basic empty weight factory specification.

Porsche 924/944/968

Summary of Special Models

Name	Factory order number	Number built	Production date	Colour	Interior
Switzerland	–	100	03/84 to 04/84	Black	Black, pinstripe
Switzerland	–	100	03/84	Zermatt Silver	Black, pinstripe
France	–	100	04/84 to 05/84	White, decals	Black, pinstripe
Celebration	M757	100 (W.Ger), 250 (US), 73 (RoW)	07/87 to 08/87	Satin Black	Black, Studio
Celebration	M757	100 (W.Ger), 250 (US), 157 (RoW)	07/87 to 09/87	Zermatt Silver	Black, Studio

M298 Prepared for unleaded fuel – manual transmission (924, 944)
M299 Prepared for unleaded fuel – automatic transmission (924, 944)
M323 Sticker: Without ECE Regulations
M325 Version for South Africa
M326 Radio: Blaupunkt Berlin IQR 87
M327 Radio: Blaupunkt Koln SQR 22
M328 Radio: Blaupunkt Bremen SQR 46 (up to 1990)
M328 Radio: Blaupunkt Symphony SQR 49 (1991 on)
M329 Radio: Blaupunkt Toronto SQR 32
M330 Radio: Blaupunkt Toronto SQR 46
M335 Automatic three-point rear seat belts
M340 Seat heating – right (944)
M341 Central locking system
M348 Forged wheels – Grand Prix White (944)
M381 Series seat, left, manual (944)
M382 Series seat, right, manual (944)
M383 Sport seat, left, electrical vertical adjustment (944)
M387 Sport seat, right, electrical vertical adjustment (944)
M389 Porsche CR stereo/manual aerial (924, 944)
M393 Forged wheel 8J/9J x 16in (944)
M394 Disc wheel, 'telephone dial' styling, 8J/9J x 16in, cast magnesium (944)
M395 light metal wheels, forged
M400 Cast aluminium wheels, 7.5J and 9J x 16in (944)
M401 Light metal wheels (944)
M409 Leather sports seats, left and right
M410 Leatherette/cloth sports seats, left and right
M411 Licence plate bracket, front (RoW)
M412 External oil cooler (944)
M414 Transmission oil cooler (944)
M416 Leather steering wheel and shift boot (924, 944)
M418 Protective side mouldings (944)
M423 Cassette container and coin box (924, 944)
M424 Automatic heating control (944)
M425 Rear wiper (924, 944)
M429 Fog lamp, white (924, 944)
M431 Leather steering wheel, 363mm (924, 944)
M432 Leather sports steering wheel, 363mm, four spokes (924, 944)
M434 Work instructions, cars for overseas (924, 944)
M437 Comfort seat, left
M438 Comfort seat, right
M441 Fader, antenna booster, four speakers (944)
M447 Emergency wheel, steel, with collapsible tyre (924, 944)
M449 Emergency wheel, forged, with collapsible tyre, 1989 on (944)
M451 Prepared for Sport group (944)
M454 Automatic speed control
M455 Wheel locks (924, 944)
M456 Sport shock absorbers and stabilisers (924, 944)
M458 16in alloy wheels (944)
M462 Sekuriflex windscreen (944)
M463 Clear windscreen
M464 Without compressor and tyre pressure gauge
M465 Rear fog lamp, left (944)
M466 Rear fog lamp, right (924, 944)
M467 Driver's side mirror, convex
M468 Graduated tint windscreen, green side glass (944)
M472 Additional rear apron (944)
M474 Sport shock absorbers
M475 Asbestos-free brake pads
M479 Version for Australia
M484 Symbols for controls
M484 Version for US, 1989 on

M485 Forged wheels, gold metallic (944)
M487 Connection for fog lamp with parking light
M488 Stickers in German
M489 Stickers and insignias in German
M490 Hifi sound system
M493 HiFi with eight speakers, Cabrio 1991 on (944)
M496 Prepared for Philips telephone network
M498 Without rear model designation
M499 Version for West Germany
M501 Turbo insignia
M513 Lumbar support, right seat (944)
M525 Alarm with continuous sound
M526 Cloth door panels (944)
M528 Passenger side convex mirror
M529 Outside mirror, passenger side convex, manual (924)
M533 Alarm system
M548 Fuel filler neck, unleaded fuel with flap (944)
M553 Version for US
M555 944S Cabrio
M556 944S Cabrio
M557 944S Cabrio
M560 Detachable roof
M562 Airbag, driver's side, to 1989 (944)
M562 Airbag, driver's and passenger's side, 1990 on (944)
M563 Airbag, passenger's side, 1989 on (944)
M564 Without airbag
M565 Safety steering wheel, leather
M572 Heating
M573 Air conditioning
M576 Without rear fog lamp (924, 944)
M586 Lumbar support, left seat (944)
M592 Brake fluid warning system
M593 Anti-lock braking (944)
M598 Insignia '16 Ventiler' (944)
M602 Third brake light 'high mount'
M612 Prepared for Philips telephone network (944)
M622 Specific engine parts for cars with two-valve engine (944)
M630 Equipment for Police
M637 Equipment for Club Sport (944)
M638 Turbo Cup (944)
M651 Electric windows (924, 944)
M652 Intermittent wipe
M655 Electric folding roof control (944)
M656 Manual steering
M657 Power steering (924, 944)
M666 Without lacquer preservation and chrome preservation
M673 Prepared for lead-sealed odometer
M675 Instrument cluster, technical lighting (944)
M684 One-piece rear seat (924, 944)
M685 Divided rear seat (944)
M686 Radio: Blaupunkt Ludwigsburg SQM 26
M688 Radio: Blaupunkt Boston SQM 23
M690 CD player CD10 radio
M691 CD player CD01 with radio up to 1988, CD02 with radio 1989 on
M715 Control PV-U production pilot run (944)
M719 Special model (944)
M734 PV-O series engine (non-series) (944)
M735 PV-O series transmission (non-series) (944)
M754 Turbo Cup (944)
M757 Special model 944 1988
M758 Special model 944 Turbo 1988 – Turbo S
M780 Remove safety certificate
M900 Tourist delivery

M912 Without vehicle identification plate
M930 Seat cover rear LLL (944)
M931 Seat cover rear KKK (944)
M932 Seat cover rear SKK (944)
M933 Seat cover rear SLL (944)
M934 Seat cover rear SSK (944)
M945 Seat cover rear SKK (944)
M946 Leather/leatherette seats (924)
M970 Front floor mats
M974 Luggage boot cover (924, 944)
M981 All leather lining
M983 Leather seats front and rear (944)
M985 Parts silver coloured (944)
M986 Partial leather lining (944)

From the UK launch in April 1982, the 944 was sold only as the 944 Lux, which included the following grouped factory options as standard: Panasonic CQ873 radio/cassette; electrically operated windows; electrically operated driver's door mirror; tinted glass; headlamp washers; front and rear anti-roll bars (21.5mm and 14mm); Sports steering wheel; rear window wiper. For 1983, side protective strips were added. For 1989 (ie, the 2.7-litre models), central locking, ABS, alarm system, automatic heating control and full electric control of the seats became standard on UK cars.

Colours & Interiors

1982-85
See 924 section (page 33).

1986 (charts VMA 7.85, WVK 103120)
Standard colours A1 Black, D4 Pastel Beige, G1 Guards Red, K3 Copenhagen Blue, P1 Alpine White.
Special colours L1 Zermatt Silver met, L5 Sapphire met, S2 Garnet Red met, S5 Crystal Green met, U8 Stone Grey met, W9 Graphite met, Y4 Kalahari met, Z6 Mahogany met.
Upholstery Leather/leatherette: black, brown, light grey. Pinstripe: black/white, brown/beige, light grey/white. Pinstripe flannel: anthracite, brown, light grey. 'Porsche' cloth: anthracite, brown, light grey.
Carpet Black, brown, burgundy, light grey.

1987 (charts WVK 102610, VMAP 6/86)
Standard colours A1 Black, B1 Summer Yellow, G1 Guards Red, G3 Crimson Red, K2 Azure Blue, P1 Alpine White.
Special colours F4 Nougat Brown met, Q3 Flamingo met, F5 Diamond Blue met, Q4 Maraschino Red met, L1 Zermatt Silver met, Q5 Almond met, Q1 Satin Black met, Q7 Nile Green met, Q2 Ocean Blue met, U8 Stone Grey met.
Upholstery Leather/leatherette: black, brown, burgundy, light grey. Pinstripe: black/white, brown/beige, burgundy/white, light grey/white. Pinstripe flannel: anthracite, brown, burgundy, light grey. 'Porsche' cloth: black, brown, burgundy, light grey.
Carpet Black, brown, burgundy, light grey.

1988 (Porsche Cars North America chart)
Standard colours A1 Black, G1 Guards Red, K2 Azure Blue, P1 Alpine White.
Special colours L1 Zermatt Silver met, Q1 Satin Black met, F4 Nougat Brown met, Q4 Maraschino Red met, U5 Almond met, Q7 Nile Green met, Q2 Ocean Blue met, U8 Stone Grey met.
Upholstery Leather/leatherette: black, burgundy, light grey (linen). Special leather: champagne. Pinstripe: black/white, burgundy/white, light grey (linen)/white. Pinstripe flannel: anthracite, burgundy, light grey. 'Porsche' cloth: black, burgundy, light grey.
Carpet Black, brown, burgundy, light grey.

1989 (charts VDA 7/88, WVK 102620)
Standard colours A1 Black, G1 Guards Red, K2 Azure Blue, P1 Alpine White.
Special colours L1 Zermatt Silver met, Q1 Satin Black met, C7 Baltic Blue met, U2 Light Gold met, U6 Velvet Red met, U1 Glacier Blue met, W5 Linen Grey met, U8 Stone Grey met.
Upholstery Leather/leatherette: black, blue, burgundy, linen grey. Pinstripe: black/white, blue/white, burgundy/white, linen grey/white. Studio check cloth: black, blue, burgundy, linen grey. 'Porsche' cloth: black, blue, burgundy, linen grey.
Carpet Black, brown, burgundy, light grey.

The 944 Turbo (1985-91)

The 944 Turbo was a wonderfully refined product, an outstanding performer in every department and an all-time favourite with road testers, yet its sales were disappointing. Not that this should have come as a surprise to Porsche, which had already set a precedent seven years previously when trying to understand why the equally impressive 924 Turbo was not achieving its expected sales targets. In both cases the answer was dependant on two main factors – the first being price.

At £25,311 in the UK, the Turbo cost about the same as the contemporary 911 Carrera (a situation reflected in most markets), and many thought the Turbo too highly priced in comparison with the regular 944 (£16,880). Meanwhile the 'yuppie' buyers Porsche had originally targeted were deserting the marque in droves in the late '80s, forcing the company back onto its original enthusiast customers. Yet to them, the 911 was the macho sports car, with the 944, even in its 250bhp version, very much the poor relation. And this was despite the fact that it was a fair match for its six-cylinder brother, both cars sharing identical power and unladen weight, accelerating to 60mph in the mid-5sec bracket and having top speeds of just over 160mph (257kph). Yet, because the 944

Evolution Outline

Feb 1985	220bhp eight-valve 944 Turbo announced.
Jul 1985	European and US launch.
Apr 1988	Limited edition 250bhp Turbo S, based on Turbo Cup specification.
Aug 1988	250bhp Turbo becomes mainstream model.
Jul 1990	Discontinued in US.
Aug 1990	250bhp Cabriolet launched.
Jul 1991	Discontinued.

The integrated front bumper cleaned up the front of the 944 Turbo and gave the car a suitably aggressive appearance.

Porsche 924/944/968

Turbo could never justify being in the same class, customers expected it to be priced to suit.

The second factor putting them off was uncertainty about the durability and driveability of the turbocharger. Porsche probably had more experience than any other company of high-performance turbos, while technologically the 944 was in a different league to the sister 924. Unfortunately the company seemed unable to get this message across, and with some reason. Throttle lag might have been much less obvious than on the 924, but the power curve was still peaky. Floor the accelerator of an S2 at 2500rpm and you got a smooth and immediate response. But on a Turbo you had to wait a couple of seconds for the turbo to spool up. Although experienced drivers might revel in the eventual surge of power, those not used to the lag had to go through a frustrating learning experience before they mastered the art of using the gearbox to keep the revs up when driving fast. Once that had been achieved, however, the Turbo was a brilliant performer.

Porsche had always planned a turbocharged 944 and spent four years developing it, including taking seventh place with a prototype 16-valve in the 1981 Le Mans 24 Hours. At one stage the car was even designated as the replacement for the 911, and therefore was required to be outstanding. Although the substitution never took place, the 944 Turbo was still notable as the first 'clean' Porsche, offering a standardised power output on lead-free fuel for every market.

For the 1989 model year, the original 220bhp model, after being in production for three and half years, was replaced by the 250bhp version, causing annoyance to customers who had forked out for the previous six months for the so-called 'limited edition' Turbo S with this engine. Their anger was increased by the company originally giving the impression only 1000 of these would be built, when the records showed the true number was some 1600.

Even if the Turbo might not have been a 911, its special appeal comes from being so much more refined. Whether in original 220bhp guise or later 250bhp form, it is still one of Porsche's all-time best models.

Bodyshell

The 944 Turbo weighed some 100kg (220lb) more than the regular 944, with only 16kg (35lb) of that down to the turbocharger and ancillaries – the rest was accounted for by the new model's high equipment specification.

Following significant wind tunnel development, the bodyshell received a new look, most obviously in the more slippery nose section which used the concealed bumper technology originated on the 928. This new front section, manufactured from a flexible composite of polyurethane and glass-fibre, housed integral driving/fog lights and indicators, as well as openings to feed air to the water radiator, oil cooler and front brakes. The

A car that means business: the 220bhp Turbo was able to accelerate almost as quickly as the contemporary 3.2-litre 911 Carrera and had the same top speed.

944 Turbo

The Turbo featured an attractive one-piece moulded polyurethane front bumper with lights and washers integrated into the design. US models had indicator repeaters on the sides. The magic word 'turbo' on any Porsche denotes electrifying performance – 0-60mph in under 6sec in this case.

front wings were revised to accommodate it, while the area where the wings met the sills was tidied up, with polyurethane/glass-fibre extensions to the sills themselves giving the car a deeper look from the side. These also helped underbody airflow, while a bonded-in windscreen, nearly flush at the sides and top edges, contributed another important aerodynamic detail. Combined with the various trim improvements described below, these brought the drag coefficient down to 0.33, some 0.02 less than the regular 944, resulting in a car which could achieve 162mph (261kph), a spectacular top speed for a production sports car, along with claimed best-in-class fuel consumption.

The previous 66-litre (14.5 Imperial gallons, 17.4 US gallons) steel petrol tank was replaced by a new moulded polyethylene tank holding 80 litres (17.6 Imperial gallons, 21.1 US gallons), along with a more environmentally-friendly closed loop breathing system.

From the 1986 model year the warranty against rust perforation was increased from seven to ten years – an important step forward in the company's confidence in its zinc-coating process.

Body Trim & Fittings

The most notable aspect about the Turbo was the complete absence of any unnecessary trim. Aside from essential fittings, such as 'elephant ear' external mirrors, everything outside was flush-fitted, including the previously mentioned windscreen with its integrated radio aerial and the low-profile jets for both headlamp and windscreen washers. Drag was further reduced by the smooth plastic engine undertray and an enlarged rear spoiler on the opening rear glass, along with a new rear underbody spoiler drawing air out from beneath the car at speed and helping to cool the gearbox and petrol tank in conjunction with small side skirts. Even compared with the regular 944, this meant the car being extraordinarily clean aerodynamically, contributing to the impressive top speed, while also reducing wind noise.

At the rear the 944 logo was replaced by 'turbo' in the same simple script style used on the 911 Turbo. After the initial launch there were few changes until the 1990 model year, when a new rear spoiler with an aerofoil cross-section gave the back end a more contemporary look.

Interior Trim

For the Turbo the Weissach designers completely redesigned the 944's interior. There was a new dashboard, along with new Recaro front seats which had electric height and backrest adjustment on the driver's side. The new seats could also be adjusted for rake and reach, and specified in either standard fabrics or half or full leather.

A new four-spoke, leather-covered steering wheel helped the driver to see the new instrument panel, while raising the wheel by 18mm (0.71in)

Porsche 924/944/968

and lowering the seat by 30mm (1.18in) improved both leg and head room. The centre console was revised and the door panels restyled. The removable sunroof, with electric tilt, came as standard.

As the car was designed as a grand tourer, much attention was given to reducing interior acoustic noise, although the ever-widening tyres negated some of the other improvements, with tyre noise especially noticeable on rough roads.

There were few changes of note subsequently, except for trim and cloth variations. Oddly, for cost reasons, the clutch pedal was manufactured from plastic from the 1990 model year.

Dashboard & Instruments

As the 944 Turbo was intended to be a flagship model, it was essential it was given a completely new design which would answer the many criticisms of the 924's dash and ventilation system. The new 'oval' design, with 928-style instruments replacing the old 'portholes', not only improved the driving position, but addressed the previous negative comments about uncontrolled heating, poor fresh air circulation, visibility of the instruments and location of the switchgear.

The new dash had larger fresh air vents, contributing to a claimed 35% increase in air flow, while both heating and air conditioning – where that was fitted – were revised to improve performance by 10% and 17% respectively. The new climate control system was fully automatic, operating independently of outside air temperature, driving speed or engine revs.

Yet, despite all the efforts made to provide a thoroughly up-to-date interior, the result still did not please everybody. There were still criticisms of instrument visibility and switch location, with one report, in time-honoured 'motoring hack' tradition, bemoaning the fact that the ashtray could not hold more than five cigarette butts.

Smart pinstripe flannel seat inlays, surrounded by black leatherette, give this Turbo a restrained interior. Door trim has switches for windows and mirrors neatly grouped ahead of the armrest. Vent in the door shutface is the air extraction path from the side of the rear luggage area.

944 Turbo

Unlike the 924 Turbo, the 944 version was fitted with a boost gauge, inset in the tachometer. The Turbo S was the first 944 to get driver and passenger airbags. There was no shortage of cooling air at face level with the 'oval' dash layout.

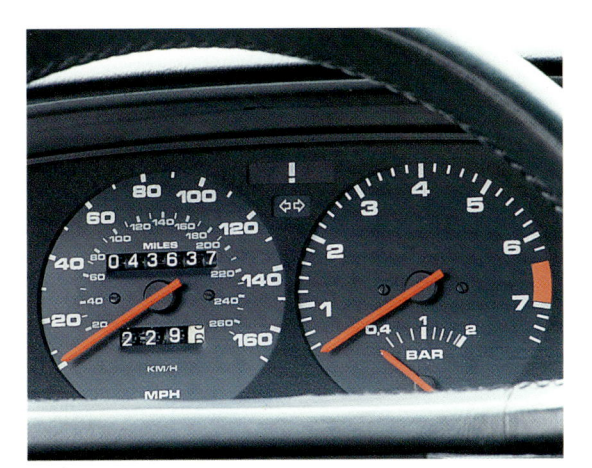

Rear luggage area of a Turbo S shows off the split rear seat backs, here in the folded position. US models were fitted with an additional brake light at the top of the glass hatch.

From the start of the 1987 model year, driver and passenger airbags became standard in US cars and optional for other left-hand drive markets, although it was pointed out that the airbag should not be considered as an alternative to the three-point seat belt, but a further enhancement of the restraint system. It was set off by a shock equivalent to hitting a concrete wall at a speed of greater than 12mph (or 20kph), with inflation occurring in only 30 milliseconds.

The driver's airbag was installed in the hub of a new steering wheel, with the passenger's mounted in the top of the dash, covered by a panel which was pushed upwards when the bag inflated. In addition the lower area of the dash was designed as an energy-absorbing knee pad to help prevent the occupants from 'submarining' out of their seats. To accommodate these changes, the glovebox on the passenger side was moved to the lower dash, where it contributed to this energy-absorbing function.

The airbag assemblies consisted of a control unit in the rear of the engine compartment and two spring/mass sensors near the door hinge posts which contained switches, together with another safety sensor mounted in the control unit, responding to frontal and oblique impacts. Porsche advised customers that the life of the airbag was a minimum of ten years, but that authorised workshop inspections should be carried out at four and eight years, and then every subsequent two years, with spaces in the service book to note this had been performed. The airbags also needed to be checked after a low-speed accident.

From February 1991 airbags became standard on all left-hand drive four-cylinder models.

Luggage Compartment

There were no major changes from the same year's regular 944.

Spacesaver tyre is less bulky than a full-size wheel, but less practical if you actually have a puncture. Shown with it are the air compressor and the bag containing the tools.

Porsche 924/944/968

The Turbo engine bay is very crowded! The two large-diameter ducts snaking around the air filter housing lead to the air-to-air intercooler.

Engine

As explained at the beginning of this chapter, the core engine (M44/51) was the first of Porsche's 'world' units, which could use lead-free fuel, whether or not a catalytic converter was fitted. This was mainly due to improved Bosch digital motor electronics, with the 'Motronic system', as it was by now called, monitoring engine speed, turbocharger pressure (restricted to a maximum of 0.76 bar), inlet manifold temperature and throttle opening.

Fuel injection was then controlled to give maximum mixture efficiency and minimal exhaust emissions, along with cold start and acceleration enrichment, a fuel cut-off on trailing throttle and a progressive rev limiter which worked by reducing fuel feed over 6500rpm. Most important of all, a knock sensor was fitted to the cylinder block between cylinders two and three, allowing the engine to run at high loading on low-octane fuel. The sensor detected the onset of engine knock and was controlled by opening the turbocharger's electrically operated waste-gate or progressively retarding the ignition. When knock was detected, the ignition retarded by 3° in the cylinder which was producing it, and then progressively advanced towards its normal setting. If the knocking persisted, the retardation could be increased further, to 6°, along with a reduction in the boost pressure.

The turbocharger installation was significantly improved from the 924 to ensure it did not suffer from the overheating problems which had characterised the earlier 924 and led, in some cases, to very short lives for both bearings and seals. The KKK K26 turbo was a new design with a water-cooled bearing housing and a thermally isolated exhaust turbine. Along with being mounted on the cooler left side of the engine, this significantly improved reliability by ensuring that operating temperatures inside the turbo never exceeded 170°C. Meanwhile a second water cooling circuit controlled temperature after shutdown by maintaining flow through the housing using a small thermostatically-controlled water pump which ran for some 30sec, followed by natural circulation. Finally, by making the

944 Turbo

exhaust gases travel further to the turbo, the temperature of the gases entering the turbine was reduced by some 90°C.

On the performance side, the lessons learned with the 924 Carrera GT were put to good use by mounting an air-to-air intercooler just ahead of the air filter housing, causing the charge air to be cooled by some 75°C before it entered the inlet manifolding, giving improved volumetric efficiency. Porsche claimed turbo lag had been all but eliminated by controlling the waste-gate electronically. Maximum boost was 1.4 bar.

As the turbocharged engine produced more heat than the unblown version, the oil pump was improved, oil volume increased to 6.5 litres and heat dissipation improved through an external oil cooler mounted in the new nose section to replace the oil/water intercooler used by the 944.

In the cylinder head, the exhaust valves were now sodium-cooled and used Nimonic seats. Valve springs were 20% harder and the exhaust ports were lined with a new ceramic material to help remove heat from the combustion chamber area. The 30°C increase in exhaust gas temperature helped the performance of the three-way catalytic converter, where this unit was fitted to the exhaust system. The pistons were forged, with the Thermally Optimised Porsche (TOP) combustion chamber design, enabling, in conjunction with the improved Motronic knock sensing, a relatively high compression ratio of 8.0:1 for lead-free and low-octane fuel.

The 1985 944 Turbo produced 220bhp at 5800rpm, with maximum torque of 330Nm (243lb ft) at 3500rpm – up a spectacular 60% on the regular 944. From the 1989 model year, the 250bhp engine then became the only unit, with all models world-wide fitted with the three-way catalytic converter. Subsequent changes were minimal, although there were improvements to the camshaft belt drive for 1990, and new fuel injection valves and a remapped Motronic programme for the 1991 model year.

Transmission

The hydraulically-operated clutch of the Turbo was strengthened by increasing the diameter to 240mm (9.49in), with action further boosted by a servo spring which reduced the pressure required on the pedal. The lining itself was asbestos-free and had a claimed greater life expectancy of the order of 60,000 miles.

The gearbox was a strengthened 016J/K, with an internal oil pump and external coil-type oil cooler, the casing including a speed sensor for the electronic speedometer. At first a limited slip differential was optional, but then became standard with the arrival of the much-improved 1988 Turbo S (or SE).

No automatic transmission option was available. Final drive was longer than that of the 944 at 3.375:1.

Electrical Equipment

The 944 Turbo was sold with a high standard of basic equipment, with central locking and air conditioning standard in many countries.

Engine equipment included a three phase 115 amp alternator and a boosted 1400 watt starter motor. Meanwhile the electrical fuses and relays were moved to a new location ahead of the firewall in the engine compartment for better cooling and access; these changes were also introduced in the mid-1985 upgrade of the regular 944.

For the 1985 cars, a new generation of radio included the Blaupunkt Hamburg SQM 24 radio/cassette with ARI (traffic advisory), with balance control and four speakers as standard. Porsche Cars Great Britain, however, chose to continue offering a choice of either the factory radio equipment or the Panasonic CQ977 radio/cassette. For the 1988 model year, the factory option changed to the Blaupunkt Ludwigsburg SQM26, or the Berlin digital radio/cassette, with UK customers offered a Panasonic CD player as a further option. These choices in turn could be boosted by an optional ten-speaker sound system. Blaupunkt kept the radios turning over regularly – for 1989 there was the Bremen SQR46, then for 1991 came the Symphony (for Rest of the World markets) or the Stuttgart (for the US). The Symphony also included RDS (Radio Data System) – on-board software which would seek out the best FM signals and identify channels. ARI (traffic advisory) was also possible.

Removable shields beneath the engine not only improved air flow under the car at higher speeds, but kept the engine bay much cleaner.

Porsche 924/944/968

Suspension & Steering

The proven 924/944 layout of MacPherson front struts with coil springs and transverse torsion bars at the rear was carried over, with the front lower wishbones and rear semi-trailing arms now made in cast aluminium for extra stiffness.

The Turbo had a firmer ride than the regular 944 with gas-filled shock absorbers and anti-roll bars with diameters of 22.5mm (0.89in) for the front and 18mm (0.71in) for the rear. Power-assisted steering, now essential with the large 205 section front tyres, was standard.

Brakes

The 220bhp Turbo had much better brakes than any previous four-cylinder Porsche, with four-piston calipers (offset to help reduce brake pad wear) on ventilated discs front and rear. Like the clutch, the pads were made from asbestos-free material world-wide from the 1989 model year. Bias front-to-rear under heavy braking was controlled by a differential valve at the rear axle, and the front brakes were additionally cooled through vents in the nose.

The most important safety feature to be introduced by Porsche on any of its models in the '80s was the Anti-block Braking System (ABS), offered as an option from the 1987 model year and made standard a year later. On a road car ABS only comes into its own in panic braking, and is at its best on wet surfaces or where grip varies from one side of the car to the other. An electrical sensor on the brake caliper senses when a wheel begins to skid and a compensation circuit built into the hydraulic system overrides the high braking force from the brake pedal, effectively reducing the pressure on the brake cylinders pushing the pads against the disc. The modifications to the steering rack and wheel hubs to fit the ABS hardware have already been described in the chapter on the 944 (see page 71).

From the start of the 1991 model year brake pad wear was monitored, and notified by an indicator in the main instrument display.

Wheels & Tyres

The first 220bhp Turbos were fitted with 205/55VR16 front and 225/50VR16 rear tyres on 7J and 8J 'telephone dial' cast alloy wheels, which in addition had anti-theft wheel locks.

For the 1989 model year the wheels changed to 928-style 'CS Design' machined finish type and, for 1990, to the 'Design 90' open seven-spoke pressure cast wheels which had first been seen on the 1988 limited edition Turbo S, in either silver finish or platinum colours.

Front suspension view shows the edge-on ventilated disc with aluminium air deflector to divert cooling air on to the disc. The Turbo introduced the significantly stiffer cast alloy lower wishbone.

944 Turbo S (1988)

One of the big competition success stories in Europe in the mid-'80s was the Porsche Turbo Cup. The series grew in stature and resulted in the Carrera Cup, which in turn grew into the F1-supporting Supercup. The first Turbo Cup races in 1986 used the new Turbo in a single model production sports car series, with improvements allowed in 1987 to give 250bhp and better handling. It was then that Porsche decided to use the Turbo Cup specification cars as a basis for a limited edition road car.

The Turbo S (known as the SE or Special Equipment in the UK) was 10% more expensive than the regular Turbo and about the same amount dearer than the current 3.2 Carrera, meaning it had to appeal to buyers wanting the car simply for its performance and exclusiveness,

The 944 Turbo's brakes were very powerful. This is the rear suspension assembly showing the four-piston caliper. ABS was introduced as an option for the 1987 model year.

944 Turbo

The 250bhp 944 Turbo S incorporated many lessons learned from the Porsche Turbo Cup race series. The warm glow of Silver Rose paint sets these limited edition models apart.

Porsche 924/944/968

and who at the same time did not think the only 'real' Porsches were air-cooled with six cylinders. Originally 1000 were planned, all intended to be painted in the same metallic silver/pink colour of Silver Rose.

From the outside the S could be distinguished by its new seven-spoke 'machined disc' 7J × 16in (front) and 9J × 16in (rear) alloy wheels carrying respectively 225/50 and 245/45 Goodyear Eagles. These big tyres provided incredible levels of grip, but at the price of transmitting more road noise into the cabin. Four-piston brakes were taken from the 928S4 and ABS was standard. Suspension changes comprised adjustable Koni shock absorbers all round, suspension bushings made from harder material, shorter (and stiffer) front springs, stiffened rear torsion bars and stiffer anti-roll bars; diameters of the bars were 30mm (1.18in) front and 20mm (0.79in) rear.

With a cue taken from the Carrera GT, there was a stylised 'Turbo' decal along the top of the right front wing, which was sneered at by purists, and a custom interior design featuring a bright striped Studio pattern in shades of pink for the seats and door inlays, over a burgundy solid colour scheme. Heated and electrically adjusted front seats were standard, as was the Blaupunkt Berlin radio/cassette with automatic peak signal seeking for a selected station. Other standard extras were split rear seat backs and a graduated tint Sekuriflex laminated windscreen.

The 2.5-litre engine received a larger turbocharger (K26-70) with a maximum boost of 1.82 bar. Although this increased throttle lag, compared to the 220bhp car, changes to the Motronic system allowed the engine to rev to 6300rpm, with the maximum power of 250bhp coming at 6000rpm. Maximum torque went up to 350Nm (258lb ft) at 4000rpm. These were very serious figures indeed for a four-cylinder engine.

The transmission was strengthened, with the torque tube diameter going up to 25.5mm (1.00in) and an improved limited slip differential fitted as standard. Meanwhile additional damping helped reduce transmission noise.

Top speed was 161mph (259kph) and 0-60mph took just 5.4sec (or 5.7sec for 0-100kph), which *Motor* magazine noted was almost identical to the 3.3-litre 911 Turbo. And while early 944 Turbos had noticeable lag on hard acceleration, the torque of the S seemed to reduce the perception of lag at low revs. The car would pull strongly from 2000rpm, the power rising without drama until it gave an electrifying shove at 4000rpm. As *Motor* noted when it compared the S with the Carrera GT, the ease and smoothness with which the power was delivered on the S made the earlier 2-litre Turbo look decidedly dated.

Inevitably, the somewhat effete pinkish silver and the Studio interior did not appeal to every customer and so, to special order, a choice could be made from the full range of colours and interiors on offer for the 220bhp model. As these different cars were not specifically identified in production, establishing how many non-silver 250bhp Turbo S models were made is difficult, although records show the factory's prediction of a limited build run of just 1000 S models actually grew to 1635.

Those buying the Turbo S (or SE) in 1988 thought they had something really special – and they did at the time – until the factory announced that this 250bhp model would become the mainstream 944 Turbo from the start of the 1989 model year. Yet the Silver Rose Turbos were indeed special – they were the first really refined water-cooled four-cylinder cars to show that there was more to Porsche than just air-cooled engines.

944 Turbo Cabriolet (1991)

Possibly the most desirable version of the 944 line, the Turbo Cabriolet, was made available as a limited edition during the summer of 1991. The new model used the running gear developed from the racing Turbo Cup 944s and the proven 2.5-

There is almost a hint of tartan in the predominantly pink 'Studio' check upholstery of the Turbo S. Rear seats had mountings for seat belts, which could also be used as anchorages for front seat shoulder harnesses, and split seat backs came as standard. Close-up shows electric adjustment for the front seats.

944 Turbo

The rare Turbo Cabriolet is probably the most desirable of all the 944 models – only 625 were made during the summer of 1991.

Porsche 924/944/968

The sumptuous leather interior of a UK-specification Turbo Cabriolet. Rear seats are for small children only, but a practical touch is the way the seat backs fold down to allow stowage of long objects from the boot area. Seven-spoke 'machine finish' wheels on the Turbo S and Turbo Cabriolet gave a new look to the model range.

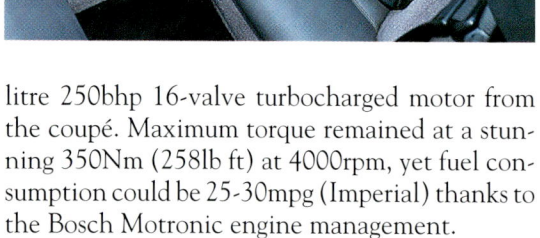

litre 250bhp 16-valve turbocharged motor from the coupé. Maximum torque remained at a stunning 350Nm (258lb ft) at 4000rpm, yet fuel consumption could be 25-30mpg (Imperial) thanks to the Bosch Motronic engine management.

The specification otherwise was as for the 1991 Turbo, excepting the Cabriolet bodyshell. For information on the Cabriolet, refer to the chapter on the 944S2 (page 97).

Like all 944 Turbos, the first driving experience of the Turbo Cabriolet centres around the fact that you can be travelling very quickly with very little effort, using a 0-100kph (0-62.5mph) time of 5.9sec and a top speed of 162mph (261kph). It is certainly a fun car in every sense of the word.

Although the original plan called for a production run of 500, 625 were made.

86

944 Turbo

Data Section

IDENTIFICATION DATA

Year	Model	Market	Chassis numbers	Engine type	Gearbox type
1985	944T	RoW	WPOZZZ95ZFN100001-0178	M44/50	016J
	944T	US	WPOAA095_FN150001-0105	M44/51	016K
1986	944T	RoW	WPOZZZ95ZGN140001-0664	M44/50	016J
	944T	RoW	WPOZZZ95ZGN100001-2760	M44/51	016J
	944T	US/Canada	WPOAA095_GN150001-7513	M44/51	016K
1987	944T	RoW	WPOZZZ95ZHN100001-1546	M44/51	016J
	944T	US	WPOAA095OHN150001-3210	M44/51	016K
	944T	Canada	WPOAA095OHN160001-0218	M44/51	016K
	944T	Cup	WPOZZZ95ZHN104101-4188	M44/52	016R
	944T	Cup/Canada	WPOAA095OHN165101-5111	M44/52	016R
1988	944T	RoW	WPOZZZ95ZJN100001-1875	M44/51	016J,R
	944T	US	WPOAA095_JN150001-1874	M44/51	016J,R
	944T	Canada	WPOAA095OJN160001-0126	M44/51	016K
	944T	Cup	WPOZZZ95ZJN104001-4094	M44/52	016R
	944T	Cup/Canada	WPOAA095OJN165001-5099	M44/52	016R
1989	944T	RoW	WPOZZZ95ZKN100001-1333	M44/52	016R
	944T	US/Canada	WPOAA295KN150001-1385	M44/52	016R
1990	944T	RoW	WPOZZZ95ZLN100001-1089	M44/52	016R
	944T	US	WPOAC295_LN150001-0144	M44/52	016R
1991	944T	RoW	WPOZZZ95ZMN100001-10250	M44/52	016R
	944T	RoW	WPOZZZ95ZMN100927-1088	M44/52	016R
	944TC	RoW	WPOZZZ95_ZMN130001-0625	M44/52	016R

Notes
No 944 Turbo (or Turbo Cabrio) production took place in Zuffenhausen (Stuttgart). The 1989 Rest of World series includes 25 Turbo Cup race cars. Production of the US 250bhp Turbo ceased in July 1990.

PRODUCTION DATA

Year	Model/market	Max power (bhp@rpm)	Max torque (Nm@rpm)	Compression ratio	Weight (kg)	Number built
1985	944T RoW	220@5800	330@3500	8.0:1	1350	178
1986	944T RoW	220@5800	330@3500	8.0:1	1350	3424
	944T US/Canada	220@5800	330@3500	8.0:1	1350	7513
1987	944T RoW	220@5800	330@3500	8.0:1	1350	1546
	944T US	220@5800	330@3500	8.0:1	1350	3210
	944T Canada	220@5800	330@3500	8.0:1	1350	218
	944T RoW Cup	220@5800	330@3500	8.0:1	1350	88
	944T Canada Cup	220@5800	330@3500	8.0:1	1350	11
1988	944T RoW	220/250[1]	330/350[1]	8.0:1	1350	1875
	944T US	220/250[1]	330/350[1]	8.0:1	1350	1874
	944T Canada	220/250[1]	330/350[1]	8.0:1	1350	126
	944T RoW Cup	250@6000	350@4000	8.0:1	1350	94
	944T Canada Cup	250@6000	350@4000	8.0:1	1350	99
1989	944T RoW	250@6000	350@4000	8.0:1	1350	1333
	944T US/Canada	250@6000	350@4000	8.0:1	1350	1385
1990	944T RoW	250@6000	350@4000	8.0:1	1350	1089
	944T Canada	250@6000	350@4000	8.0:1	1350	144
1991	944T RoW	250@6000	350@4000	8.0:1	1350	787
	944TC RoW	250@6000	350@4000	8.0:1	1350	251
Total						25,245

Notes
[1] Denotes 1988 Turbo S/SE. Production volumes are: Rest of World, 917; US, 718 (interesting numbers for race fans!). Total run of 1988 944 Turbo S: 1635.

IDENTIFICATION

Chassis numbers There was no specific sequence for the 250bhp Turbo S models, which were part of the 1988 production run, as follows: RoW, 95JN104001 to 95JN101875; US, 95JN150001 to 95JN151874. However, 1988 Turbo S cars are distinguishable by their engine numbers. The 250bhp engine is M44/52, with serial numbers starting 47J00001-1830 (that is there were 1830 250bhp engines made during the 1988 model year).

Engine numbers The 220bhp engines were the early RoW M44/50 (discontinued after 1986), with a serial number that started 44, followed by the year letter (eg, F for 1985) and the build number. For the US and later world spec engine M44/51, the serial started 45 followed by the year letter and number (eg, 45 H00083 etc, for a 1987 build engine). From the 1989 model year, the 250bhp engine serial numbers began 47, followed by the model year designator: 1989, K; 1990, L; etc.

KEY FACTS

Fuel European version: 95 RON unleaded; 91 RON unleaded possible for limited use, with loss of power.
Dimensions Length, 4230mm; width, 1735mm; height, 1275mm; front track, 1477mm; rear track, 1451mm; wheelbase, 2400mm; ground clearance (empty), 120mm; turning circle, 10.3m; kerb weight, 1350kg (1985 Europe, RoW).
Capacities Fuel tank, 80 litres; engine oil, 6.5 litres; coolant, 8.5 litres; transmission (manual), 2.2 litres; screen/headlamp wash, 6.0 litres.

OPTIONS
See 944 section (pages 73-74).

COLOURS & INTERIORS
See 944 and S2 sections (pages 74 and 99).

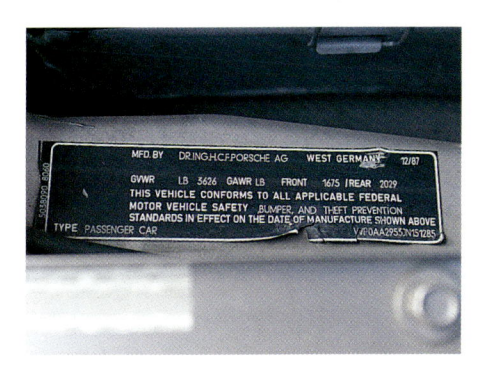

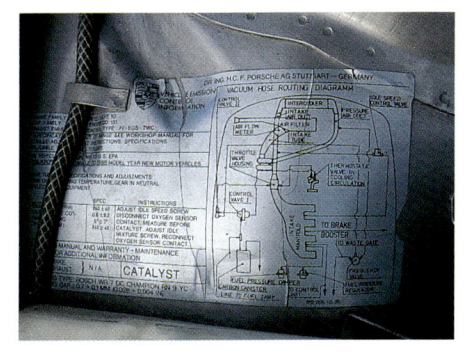

Two labels applied to US models, giving safety and emission control information.

Porsche 924/944/968

The 944S & S2 (1986-91)

The 944S is a bit of a puzzle. On paper, the prospect of a 16-valve version of the 2.5-litre engine was extremely attractive, especially as the 16-valve head had its origins in the 924 GTP car which finished seventh at Le Mans in 1981. Yet this head did not find its way into a production car until late 1986 and then only finally appeared to disappoint.

By then both the regular 944 and the Turbo were established and the market was already over-heating, with Japanese competition aggressively attacking Porsche's market share to the point that in 1987 the 944 had serious rivals even in its major market of the US. Worse still, both the market and exchange rates were about to go against all foreign importers, squeezing luxury car makers like Porsche into an almost impossibly small corner. The company reacted by trying to take a

Evolution Outline

Sep 1985	Frankfurt Show design study reveals prototype 16-valve engine.
Aug 1986	Production starts of 2.5-litre 944S.
Jan 1989	Production starts of 3-litre 944S2.
Jul 1989	Production starts of 3-litre 944S2 Cabriolet.
Jan 1991	Production of four-cylinder 944 piloted in Zuffenhausen.
Jul 1991	944S2 production ceases.

The 190bhp 944S introduced the 16-valve head to the 944 range. The most notable feature was a delicious twin-cam roar under hard acceleration. Black 211bhp 944S2 (facing page) reflects an almost surreal potency in this gleaming front view. Seven-spoke 'Design 90' wheels were fitted from the 1990 model year, and wide 225/55ZR16 tyres immediately distinguish the 944S2 from the earlier 944S.

88

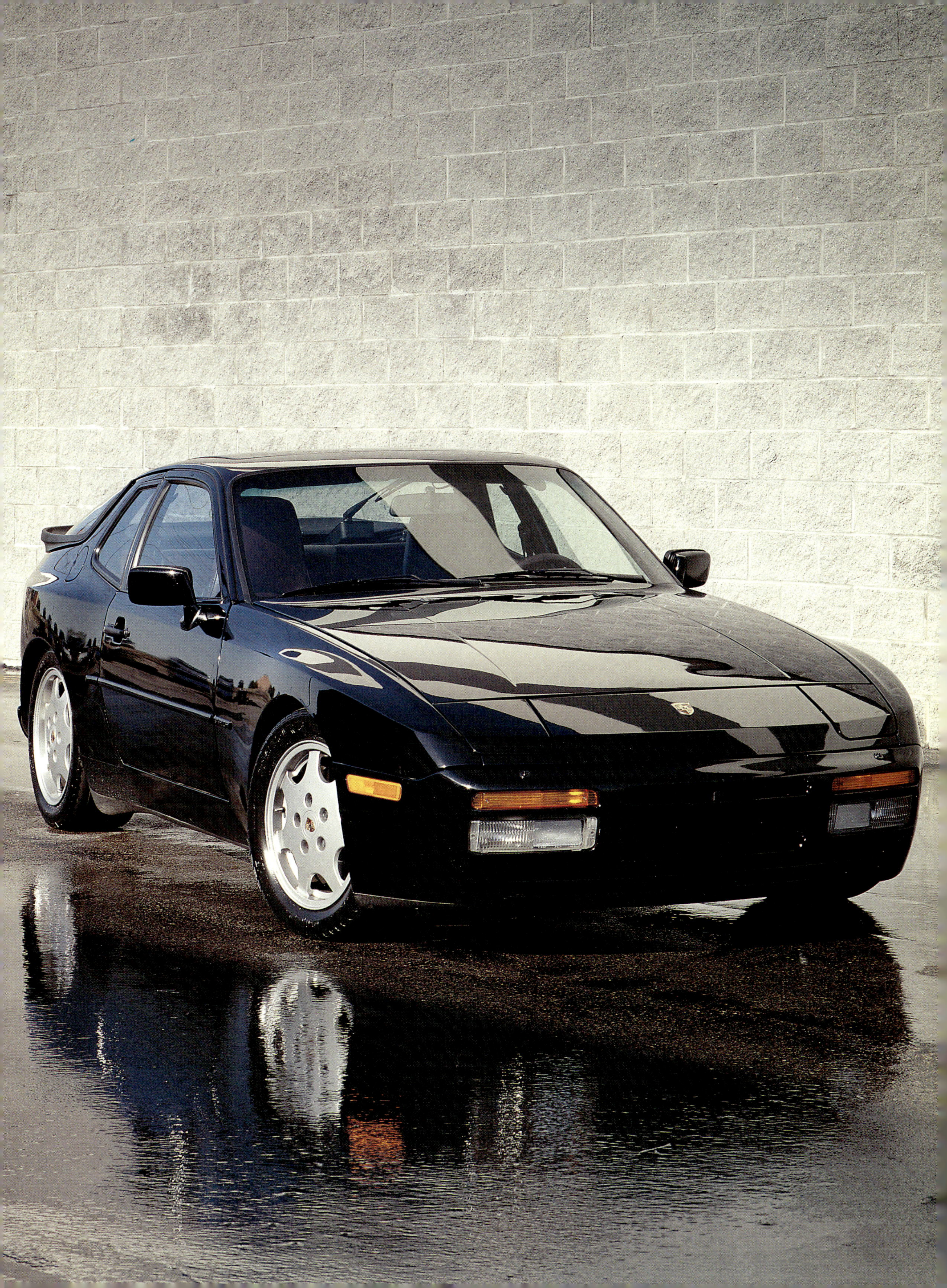

Porsche 924/944/968

larger slice through rolling out the 190bhp 944S, priced between the regular 944 and the Turbo, but in a steadily contracting market this policy was inevitably doomed.

In July 1989, Porsche had celebrated the completion of the 300,000th water-cooled four-cylinder car at Neckarsulm over a time span of 14 years. Yet the market was already turning its back on the model that had made Porsche a wealthy company. For the 944 series, although by now the four-cylinder family was stronger than ever, the writing was on the wall.

On the positive side, however, those lean years have left enthusiasts with a product which was made by a company pumping in all its expertise simply to survive. Although the 944S may not at first have been a shining star, the S2, and particularly the Cabriolet which followed, rank among Porsche's best and were worthy flagships at such a testing time.

The Cabriolet completely redrew the possibilities for a small car from the company, being much more successful than the 911 Cabriolet from a styling viewpoint. Being front-engined, it had no ugly rump behind the occupants and it looked just

Note the rear underbody spoiler, first seen on the Turbo (and often retro-fitted by owners to other models), and the 'quartered' rear bumpers found on US models.

The January 1989 evolution from 944S to 944S2 saw engine capacity rise from 2.5 litres to 3 litres – and also brought a different graphic style for tail badging.

944S & S2

duction, which was to be wholly in-house from the start of the 1991 model year. The records show that a mere 549 944S2s (Coupés and Cabrios) were built, those with the more desirable S for Stuttgart (rather than N for Neckarsulm) in their chassis numbers nowadays carrying a premium among enthusiasts. In addition, because dealers were unloading unsold 944s during 1991 and early 1992, there will always be some excellent bargains on these cars, as their prices have never recovered from the initial discounting.

Bodyshell

The 944S adopted the uprated mid-1985 944 bodyshell which benefited from the significant improvements developed for the new Turbo model the previous year. And because it retained the 944's separate front bumper and front lower panel, it was not until you got up close that you could tell the difference.

For 1989, the new S2 adopted the lower drag, full Turbo bodywork, contributing to the revised model's impressive top speed of 150mph (241kph).

Body Trim & Fittings

Apart from the small '944S' logo on the rear and extremely discreet '16 Ventiler' badges ahead of each door, there were no external visual identifiers for these models.

The 1988 models were easy to distinguish as they were given the Turbo's rear underbody spoiler, while still retaining the regular car's front bumper. The 1991 S2 models were distinguished by the Turbo's wing spoiler, finished in body colour, which significantly improved visibility when reversing.

Interior Trim

The interior trim of the 944S was standardised with the same year's 944 model until the start of the 1989 model year, when it was aligned with the more sophisticated Turbo.

With the immediate success of Turbo Cup events in Europe, 944S owners could also specify a special Club Sport option, which deleted most of the unnecessary electric options, including air conditioning and electric operation for windows and seats, as well as the underbody PVC sound deadening. Additional options were sports (bucket-type) seats, six-point safety harnesses and a roll cage.

Airbags for both driver and passenger on US models were an important new option for normally-aspirated 944s for the 1987 model year. For 1991 automatic three-point seat belts were fitted to the rear seats.

The S2 adopted the cleaner front bumper panel of the Turbo. Indicator repeaters on the flanks were a feature of US models. A discreet distinguishing feature of all S models was a small '16 Ventiler' (16 Valves) badge in front of each door.

as good with the cleverly-styled convertible top up or down, which cannot be said of the 911. One can only wonder how much its style and appeal inspired Styling Director Harm Lagaay and his team when they were conceiving the Boxster prototype, first shown in 1993.

On 1 January 1991 Porsche announced it was transferring four-cylinder production from Neckarsulm to the new Werk V building in Zuffenhausen in readiness for the start of 968 pro-

Porsche 924/944/968

Dashboard & Instruments

There were no differences between the 944S and the 944. For the S2 the tachometer did not include the boost gauge used on the Turbo.

Luggage Compartment

There were no differences compared to the 944 (944S) or the later 944 Turbo (944S2).

Engine

Apart from enlarged inlet ports, the 16-valve cylinder head for the 944S engine was taken directly from the new 5-litre V8 of the 928S. The block came from the standard 944, retaining the same bore and stroke of 100mm (3.94in) × 78.9mm (3.11in) as its smaller brother, the capacity as a result remaining at 2479cc (151.3cu in) – exactly half that of the new 928S4. Power and torque respectively peaked at 190bhp and 230Nm (170lb ft), while maximum revs went up to 6800rpm from the eight-valve's 6500rpm.

The exhaust camshaft was driven from the crankshaft by a single toothed belt, which was stronger than the one previously used on 944 models and also had a new automatic tensioner that was to find its way on to other models in due course. Drive to the intake camshaft was achieved using a centrally mounted Simplex chain. The valves were arranged at an included angle of 27.5° in a pent-roofed combustion chamber. Each of the four valves per cylinder was 35mm (1.38in) in diameter, 3mm (0.19in) smaller than in the two-valve head. Vertical arrangement of the spark plug between the valves was an important point for

Electrically operated front seats in this 944S2 feature impressive side restraint, while view of rear seats shows lap and diagonal belts and speaker location. Trim is Burgundy leather.

944S & S2

The 944S dashboard, seen on a car with Burgundy red pinstripe upholstery, was identical to that of the contemporary 944. The bulkier airbag steering wheel (below), which became standard on US cars for the 1990 model year, did not improve the driver's view of the controls.

improved flame propagation. Magnesium was used for the new wider cam cover, the intake manifolding and the distributor drive housing, while the exhaust was modified for the new head layout and there was a new sump casting.

One of the engineers' justifications for developing the 16-valve head had been improved exhaust emissions and greater fuel efficiency, with lower grade fuel possible because of the improvements in the engine's ability to breathe. Yet to achieve these goals fully, Porsche needed further help from Bosch, whose revised Motronic engine management now included two knock sensors – one more than in the Turbo – and self-adaptive Lambda control. Although fuel pressure was increased, a new closed-loop fuel cut-off system reduced the amount lost through vaporisation.

The Motronic system was enhanced for 1988 so that it now stored diagnostic information about the injection, ignition and knock control which could be read off when servicing. All models, with or without catalytic converter, had standardised exhaust systems and for 1989 were given the three-way converter as standard, along with the Lambda probe and active carbon tank ventilation.

For the 1989 model year the engine was enlarged to 2990cc (182.5cu in) for the new S2, maximum power and torque increasing to 211bhp

Porsche 924/944/968

The rear luggage area of a US-specification 944S2. Note the preference for the single rear seat back and roller blind cover, which is seen retracted and extended. In the extended position, it is difficult to tell there is a luggage area below with the rear hatch closed.

and 280Nm (206lb ft). Shared with the 2.7-litre 944 introduced at the same time, the new aluminium block had cylinder spacing that allowed the bore to be increased to 104mm (4.09in), but for the S2 the stroke was also increased to 88mm (3.46in), this being accommodated by using shorter skirts on the pistons.

The new block was lighter and stiffer, and used cooling technology which had been developed during the TAG Formula 1 programme. As well as using siamesed cylinders, the coolant volume in the block was reduced from 1.1 litres to 0.55 litres, a development that confirmed that almost no

The tool kit of the 944S was shared with the 944 and included everything for basic work on the car.

944S & S2

Comparison between the two capacities of 16-valve engine: 2479cc/190bhp (facing page) and 2990cc/211bhp (below). The most noticeable difference is the revised inlet manifolding of the longer-stroke engine, which used the lighter, stiffer block from the 2.7-litre 944 and featured technology from the TAG F1 programme. Although the camshaft cover was made from magnesium alloy, the 16-valve engine was notably heavier than the eight-valve version.

supplementary cooling was required below the level of the combustion chambers. Formula 1 knowledge also resulted in lower oil capacity and less foaming of the oil. The engine also used tighter dimensional tolerancing for its components and there was a new plastic sump pan. Yet again the Motronic system was upgraded and there was a new quadrilateral intake manifold which encouraged resonant charging of the cylinders, giving almost a turbo effect without the turbo.

This increase in capacity transformed the 16-valve engine, previously criticised for its lack of low to mid-range torque. Developments continued for the 1990 model year with improved engine management, automatic three-sensor knock control, a larger light alloy external oil cooler and an engine compartment ventilation system with a three-stage fan.

Transmission

The 944S used the strengthened gearbox and driveshafts (by Oerlikon) adopted for the 220bhp Turbo. Fifth gear on the S2 was longer than on the original S, as was the final drive, up to 3.875:1 from 3.889:1, reflecting the improved torque of the 3-litre engine. Although there was the option of a limited slip differential, there was no automatic transmission, which disappeared with the 2.7-litre 944.

Electrical Equipment

There were no differences between the electrical equipment of the 1987 944 and the new S. Both the S and the 1989 S2 used a 115 amp/1610 watt alternator with a 63Ah battery.

The 1990 model year introduced a time delay before the interior light turned off, while for 1991 the previously optional combined central locking and alarm system was specified as standard from the factory. The height of the headlamps could be electrically adjusted from the driver's seat.

Suspension & Steering

For the S2, the anti-roll bars were stiffened to 26.8mm (1.06in) front and 16mm (0.63in) rear. Optional sports suspension included the package found on the Turbo S, with 30mm (1.18in) front and 20mm (0.79in) rear anti-roll bars. These dimensions were some considerable way from the

Porsche 924/944/968

944S & S2

The convertible roof line, the flattened rear deck and the revised curve of the side window gave a completely new appearance to the 944. The cut line for the conversion can be seen extending back horizontally from the door at waist level. A tonneau ensured that all the untidy parts were hidden when the soft-top was down.

A rocker switch on the centre console operated the powered soft-top.

1982 944's 20mm (0.79in) front and optional 14mm (0.55in) rear, showing the progress made in suspension design and ride comfort.

Brakes

The braking system of the S was identical to that of the 1987 944. New for 1987 had been the brake pressure regulator on the rear brakes, also fitted to the earlier Turbo and changing the brake bias from the rear wheels to assist in preventing lock-up. ABS was a new option for 1987 and became standard ex-factory for 1991, although many markets had already been offering it as a matter of course.

The S2 adopted the 250bhp Turbo's braking system, itself derived from that of the 928 S4.

Wheels & Tyres

The same 'telephone dial' 7J × 15in wheels were fitted to the 944S as to the regular 944, with optional 7J × 16in front and 8J × 16in rear forged disc style wheels. From the start of the 1987 model year another option was platinum, rather than silver, centres for the optional wheels. The tyres were 195/65VR15s all round, with optional 205/55VR16 front and 225/50VR16 rear.

For the 1989 S2, the standard fitting was 7J × 16in and 8J × 16in 928-style pressure-cast light alloy wheels, called 'CS Design' by Porsche. For 1990 these changed to the seven-spoke 'Design 90' wheels, first seen on the 1988 Turbo S. Tyres were 205/55ZR16 front and 225/50ZR16 rear. Locking wheel nuts were factory specified from the start of the 1991 model year.

944S2 Cabriolet (1990-91)

The Cabriolet was not really a special model as such, but it is worth a separate entry as it was produced in such relatively small numbers over a period of just two years.

The Cabrio Studie, a one-off design study shown at the 1985 Frankfurt Motor Show, had been produced by Bauer, but controversially the production contract to convert standard coupé bodyshells and build the convertible tops went to the American Sunroof Company. ASC then had to start a green field factory in Heilbronn and overcome various problems, most importantly significant chassis flexing when the coupé's roof was removed. The Cabriolet was alleged to have been up to two years late reaching production because of these delays.

After much work the empty weight of the Cabriolet finished up some 70kg (154lb) more than the coupé due to its additional undertray, cross-member sections and stronger door pillars.

Porsche 924/944/968

The strengthening round the sill area is noticeable because of the reduced access to the handbrake. The windscreen was steeply raked with its top edge 60mm (2.4in) lower than the coupé's, while at the rear the opening glass was replaced by a conventional boot lid. Unfortunately the rear luggage compartment was reduced to a fraction of its previous capacity, with the deep wells behind the rear wheels closed off. At least the vestigial rear seat backs still folded flat, but the lowered soft-top took up most of the rear seat space.

From the outside, however, the styling of the new model was an outstanding success. The result was an extremely attractive machine which had lost the rapidly dating side aspect of the coupé, particularly noticeable in the rear windows, thanks to a completely new roof line and a 'notchback' tail.

Getting into a Cabriolet, especially for the taller driver, does require agility to thread your legs past the low-set steering wheel and the steeply-raked windscreen. The trick, as owners soon discovered, was to slide the seat back before you got out of the car. In addition, with the top up, all-round visibility was poor, while the high front seat backs made visibility for reversing difficult, even with the roof down.

The Cabriolet benefited from a rationalisation of the factory parts inventory from 1990 by using – along with the closed S2 – the Turbo's bodywork and running gear, which gave a firm and confident ride. Inside, air conditioning was standard, along with electric operation of the soft-top after it had been manually released from the windscreen with two detachable levers.

The Cabriolet shared the S2's brilliant 211bhp 16-valve 3-litre engine, and a year later a limited edition received the 2.5-litre turbocharged engine (see 944 Turbo chapter, pages 84-86). Having a shorter windscreen than the coupé ought to have yielded a top speed advantage, but 149mph (240kph) was quoted for both open and closed models, with the Cabriolet slightly slower on acceleration to 60mph due to the extra weight of the body strengthening.

As an all-rounder, it might have lost much of practicality of the coupé, but it was, and still is, *the* car to have on those warm, summer days when the road ahead is long and winding.

944SE (1991)

With the 968 first coming along at the end of 1991 and moving into other markets during 1992, dealers sought to unload their stocks of slow-moving 944s, giving some splendid deals. One special offered in the UK was the 944SE, at a price premium of only £2000 on the standard S2.

The 944SE had full Sport suspension, with ride height lowered by 30mm (1.2in), and engine power increased to 225bhp. Aesthetically, it was finished off with a Porsche Sport steering wheel from the *Exclusiv* programme, a rear spoiler colour-coded to the body, and custom 'SE' decals on the flanks.

Boot space – and convenience of access – was dramatically reduced in the Cabriolet. Here the rear seat backs are dropped, but whether a golf bag would go in is questionable.

The side windows were revised to the new line of the Cabriolet roof and there was extra strengthening around the door pillars. This car has the smart black pinstripe upholstery. The adhesive-backed stone guard ahead of the rear wheels helped prevent stone chips.

944S & S2

Data Section

IDENTIFICATION

Chassis numbers The 17-character VIN chassis numbering system is fully explained in the 924 chapter (see pages 31-32). The 1989 944S2 had the same VIN number as that year's 944. There were no US 944 S2s in the 1989 model year. From 1 January 1991, pilot builds of 944S2s transferred to Zuffenhausen from Neckarsulm, and 11th digit of chassis number changed from N to S (denoting Stuttgart). Records show that all were for Rest of World markets and that 130 coupés and 419 Cabrios were built in Stuttgart.

Gearbox numbers Serial number for 944S (083D five-speed) began AGP followed by date of build (ddmmyy). If a limited slip differential is fitted (option M220) serial number starts AGR followed by date of build (ddmmyy).

Engine numbers Year of engine manufacture can be identified by its serial number. The M44/40 2.5-litre S engine was designated 42 followed by H or J (depending on whether it was built in '87 or '88) followed by the serial number. The 3-litre engine (M44/41) was also designated 42 followed by the year letter (K, L or M for '89, '90 or '91) and the build number.

KEY FACTS

Fuel European version: 95 RON unleaded.
Dimensions Length, 4230mm; width, 1735mm; height, 1275mm; front track, 1472mm; rear track, 1451mm; wheelbase, 2400mm; ground clearance (empty), 125mm; turning circle, 10.75m; kerb weight, 1340kg (coupé) or 1390kg (Cabrio).
Capacities Fuel tank, 80 litres; engine oil, 7 litres; coolant, 8 litres; transmission (manual), 2 litres; screen/headlamp wash, 6 litres.

OPTIONS

See 944 section (pages 73-74).

COLOURS & INTERIORS

1986-89
See 944 section (page 74).

1990 (chart WVK 102020)
Standard colours A1 Black, P1 Alpine White, G4 Guards Red, K2 Azure Blue.
Special colours C6 Crystal Silver met, S6 Cyclamen Red met, C7 Baltic Blue met, Z6 Titanium met, U1 Glacier Blue met, U6 Velvet Red met, L7 Panther Black met, W5 Linen Grey met.
Upholstery Leather/leatherette: black, blue, burgundy, linen grey. Pinstripe velour: black/white, blue/white, burgundy/white, linen grey/white. Studio check cloth: black, blue, burgundy, linen grey. 'Porsche' cloth: black, blue, burgundy, linen grey.
Carpet Black, blue, burgundy, linen grey.
Cabrio soft-top Black, blue, burgundy.

1991 (Porsche Cars Great Britain chart)
Standard colours A1 Black, F2 Maritime Blue, P1 Alpine White, G4 Guards Red, K2 Azure Blue, G4 Rubystone Red.
Special colours C6 Crystal Silver met, S6 Cyclamen Red met, F6 Cobalt Blue met, Z6 Titanium met, U1 Glacier Blue met, L7 Panther Black met.
Upholstery Leather/leatherette: black, cobalt blue, classic grey. Studio check cloth: black, cobalt blue, classic grey. 'Porsche' cloth: black, cobalt blue, classic grey.
Carpet Black, cobalt blue, classic grey.
Cabrio soft-top Black, cobalt blue, classic grey.

The VIN number was displayed behind the windscreen; central locking and an alarm were standard factory fit from 1991.

IDENTIFICATION DATA

Year	Model	Market	Chassis numbers	Engine type	Gearbox type
1987	944S	RoW	WP0ZZZ94ZHN40001-2635	M44/40	083D
	944S	US/Canada	WP0AA094_HN450001-3127	M44/40	083D
	944S airbag	US	WP0AA094_HN465101-5113	M44/40	083D
	944S Sport	US	WP0AA094_HN480101-0102	M44/40	083D
1988	944S	RoW	WP0ZZZ94ZJN40001-1305	M44/40	083D
	944S	US	WP0AA094_JN460001-5561	M44/40	083D
	944S	Canada	WP0AA094_JN450001-0188	M44/40	083D
1989	944S2	RoW	WP0ZZZ94ZKN402731-4941	M44/41	083F
	944S2	US	WP0AB294_KN450001-2691	M44/41	083F
1990	944S2	RoW	WP0ZZZ94ZLN400001-2872	M44/41	083F
	944S2C	RoW	WP0ZZZ94ZLN430001-2114	M44/41	083F
	944S2	US/Canada	WP0AB294_LN450001-0449	M44/41	083F
	944S2C	US/Canada	WP0CB294_LN48001-1824	M44/41	083F
1991	944S2	RoW	WP0ZZZ94ZMN400001-2608	M44/41	G83F
	944S2	RoW	WP0ZZZ94ZMS400001-0130	M44/41	G83F
	944S2C	RoW	WP0ZZZ94ZMN430001-1140	M44/41	G83F
	944S2C	RoW	WP0ZZZ94ZMS430001-0419	M44/41	G83F
	944S2	US	WP0AB294_MN410001-0510	M44/41	G83F
	944S2C	US	WP0CB294_MN440001-0562	M44/41	G83F

PRODUCTION DATA

Year	Model/market	Max power (bhp@rpm)	Max torque (Nm@rpm)	Compression ratio	Weight (kg)	Number built
1987	944S RoW	190@6000	230@4300	10.9:1	1280	2635
	944S US/Canada	190@6000	230@4300	10.9:1	1280	3127
	944S airbag US	190@6000	230@4300	10.9:1	1280	13
	944S Sport US	190@6000	230@4300	10.9:1	1280	2
1988	944S RoW	190@6000	230@4300	10.9:1	1280	1305
	944S US	190@6000	230@4300	10.9:1	1280	5561
	944S Canada	190@6000	230@4300	10.9:1	1280	188
1989	944S2 RoW	211@5800	280@4000	10.9:1	1310	4941
	944S2 US/Canada	211@5800	280@4000	10.9:1	1310	2691
1990	944S2 RoW	211@5800	280@4000	10.9:1	1310	2872
	944S2C RoW	211@5800	280@4000	10.9:1	1380	2114
	944S2 US/Canada	211@5800	280@4000	10.9:1	1310	449
	944S2C US/Canada	211@5800	280@4000	10.9:1	1380	1824
1991	944S2 RoW	211@5800	280@4000	10.9:1	1310	2738
	944S2C RoW	211@5800	280@4000	10.9:1	1380	1559
	944S2 US	211@5800	280@4000	10.9:1	1310	510
	944S2C US	211@5800	280@4000	10.9:1	1380	562
Total						33,091

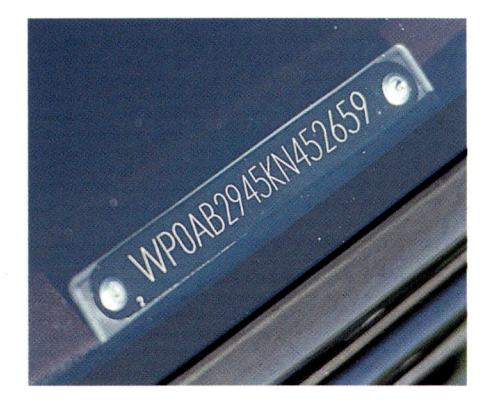

Porsche 924/944/968

The 968 (1991-95)

Porsche was in turmoil during the early '90s. Undignified changes of Chief Executive Officers mirrored the way that even racing activity was all at sea. There was the Porsche-engined Arrows Formula 1 programme, with the sight for millions of TV viewers of a car broken clean in two at Monaco, and the Indycar projects, with a wheel being similarly viewed bouncing along the pit lane at Indianapolis. For a company which had always let its racing efforts speak for it, its performance on the track in the early '90s showed how far it had lost direction.

In a market which was bad anyway, the company had desperately needed new models since 1988, and the new decade began with poor sales of the various 944 updates. Prevented by financial constraints from completely replacing the series, Porsche had already taken the decision to develop what emerged as a vehicle claimed, in the curious language of modern marketing-speak, to be 83% 'new'. This model, the 968, was still clearly derived from the outgoing 944, but re-

Evolution Outline

Aug 1991	968 launched (right-hand drive versions May 1992).
Dec 1992	968 Club Sport announced.
Feb 1993	968 Turbo S and RS announced.
Jan 1994	968 Sport announced for UK market.
Jul 1995	968 discontinued.

The 968 was dramatically removed from the first 924 in terms of technology, yet retained the inherent excellence of the original design.

100

968

styled and re-engineered to keep Porsche ahead in the 3-litre sports car class, with the four-cylinder 944S2 engine significantly improved by a new variable valve timing design and other updates intended to yield better power and torque, together with reduced emissions.

'Evolution, rather than revolution,' claimed Paul Hensler, then Director of Group Development at Porsche.

Yet, sadly for the company, press reaction to the car's launch in August 1991 was not just 'underwhelming', but positively negative. The 968's real problem was that it lacked flair, the reviewers complained. It did not have the snappy image or charisma to fire their imaginations, and therefore, *per se*, they pronounced, those of potential customers.

The only model which sparked any strongly favourable comments was the Cabriolet. With its elegant, and still fresh, convertible roof line, the new rounded front and rear seemed to fit far more acceptably than the 15-year-old styling that still characterised the coupé's mid-section.

A year after launch, in confirmation of the motoring writers' judgement, stories were coming out of Zuffenhausen that the production line in Werk 5 was only producing a small number of cars. To try to counter this, Porsche rushed the 968 Club Sport into production for 1993 and loaded mainstream 968s with quality options as standard. But despite these huge efforts to keep the model going until the Boxster arrived in the autumn of 1996, production finally ground to a permanent halt in the summer of 1995.

The four-cylinder water-cooled generation, lasting from 1976 to 1995, had come to an end, leaving a legacy of fine cars, together making up well over a third of Porsche's total lifetime production. And enthusiasts should not be put off in the slightest by the early negative comments. The 968 models still represent the very best of the four-cylinder water-cooled Porsches.

Bodyshell

Externally, the overall shape was brought into line with the identity which is now considered wholly Porsche. In particular, the new details introduced to the coupé and Cabriolet included the trapezoidal shape of the bonnet and mirrors, the 928-type exposed headlamps and the inlet openings at the front.

The lines of the car were softened, both metaphorically and literally, by using flexible polyurethane front and rear panels. Drag coefficient came out at 0.34, slightly worse than the

The rear end was significantly revised, with new rear wings and a moulded one-piece tail section made from flexible polyurethane. Note the US-specification rubber buffers either side of the number plate.

Porsche 924/944/968

The 968 Cabriolet continued the appeal of wind-in-the-hair motoring in the four-cylinder range, and some enthusiasts rated it the most attractive open sports car Porsche had ever made. The only negative was the loss of boot space.

968

The rear wing (near right) was carried over from the outgoing 944 and complemented the 968's new style. New pop-up headlamps (far right) – the exposed headlamp was the new Porsche family look for the '90s – were surrounded by soft curves. The rubber buffer, part of US specification, reduces the forward-facing area of the indicator light compared with the European model.

Rear light units were bonded into the polyurethane moulding. Considerable styling effort went into softening the sharp lines of the car – as seen here on the radiused corners of the light assemblies.

S2/Turbo's 0.33, partly because of the slight raising of the front wing line to accommodate new pop-up headlamps. At the car's launch it was announced that the original 924/944 pop-ups, folding forwards out of sight, had been so copied by Japanese manufacturers that the company had felt a distinctive change was needed. That particular goal was now achieved through having the lamps retracting backwards, with the lens flush with the wing, although whether this was also a stylistic improvement was the subject of much debate among Porsche enthusiasts.

But if the front end treatment received a mixed reception, the rear was universally acclaimed for its neatness. The previous rear panel, with its separate bumper and underbody spoiler, had gone, with the light clusters now integrated into the new panel.

There was also a new range of colours which suggested a return to some of the wild options last seen in the early '80s. Finally Porsche was seen to be trying to shake off the Guards Red image which had taken over the model. Meanwhile, for 1993, concern for the environment took a step forward when water-based paints were used for some of the solid colours.

But, for all the new development, the motoring reviewers had a point. The car was still beset with its '70s-vintage 7mm gaps between panels, most noticeably around the doors and at the sides of the bonnet, where the gap was so large the wing attachment bolts were clearly on view. These gaps had stayed because of the economic need to carry over the chassis tooling from the 944, but in 1991, when the industry standard had become 3-4mm, visible bolts were a poorly considered detail which said a lot.

Body Trim & Fittings

The new car followed the contemporary fashion for minimum additions, using few of the external trim details from the 944, while the new main sections for the nose and rear were the result of the effort to give the car a brand new appearance without major changes in tooling. A new flexible moulding below each door also reduced the visual profile from the side, while the rear three-quarter glass was now bonded into place.

Porsche 924/944/968

The tailgate on the coupé, complete with wing spoiler, was unchanged from the 944, but the external mirrors and door handles were new. The previous 'elephant ear' door mirrors were replaced by attractive, more streamlined units, still electrically adjustable and heated, which took their cue from the 911 Turbo. The door handles, previously VW/Audi units, were replaced by more stylish, and now colour-coded, items.

The appearance of the engine when the bonnet was opened was also much improved. In the '90s, as owners adjusted to looking at their engines, rather than touching them, deliberate trimming of the engine compartment became a phenomenon Porsche was happy to follow. The aim in all cases was a perception of neatness and order, inferring quality and reliability, even at the price of reducing accessibility for mechanics actually servicing and working on the cars.

The cylinder head and inlet tract castings were tidied up, and black plastic mouldings introduced at the front and rear of the engine compartment. The new camshaft cover was sculpted to blend neatly into a new casting which covered over the fuel injection pipework and spark plug leads to the extent that the earlier clutter was replaced by a view of no pipework at all over the top of the engine. At the firewall, previously a very busy area containing brake and clutch ancillaries, wiring and other plumbing, a smart vacuum-formed moulding covered all. Similarly, at the front, another moulding wrapped itself round the air filter and fuel injection hardware, even incorporating a neat compartment at lower right to house work gloves.

In an environment-friendly touch, identification symbols were also placed on the plastic containers to help with their recycling at the end of their natural lives.

Interior Trim

Inside the new car, the trim colours and materials were standardised with the 928 and 911, and a new range of materials and colours was offered. A leatherette interior was standard, with 'Porsche' motif cloth for the seat inserts. As before, the front seats had electric height adjustment.

An airbag, integrated into a bulbous new steering wheel, was standard on the driver's side for all left-hand drive markets. On the passenger side it was enclosed behind a blow-out panel above the glove compartment, as on the 944. Right-hand drive models did not get a passenger airbag until a year later, for the 1993 model year, but in the meantime received a chunky new steering wheel which was a joy to use.

With environmental compatibility playing a greater part in design, the car used CFC-free refrigerant (R134a) for its factory-optional air conditioning. For 1994 the ventilation system was upgraded with two particle filters in the fresh air intakes, each of which could stop particles down to a size of 1 micron (0.0001mm), with a stated change interval every 12,000 miles (20,000kms). The same model year also brought a new seating generation, with leather and full electronic adjustment for the front seats.

The 968's large lower air intake provided cooling to the front brakes as well as the radiator.

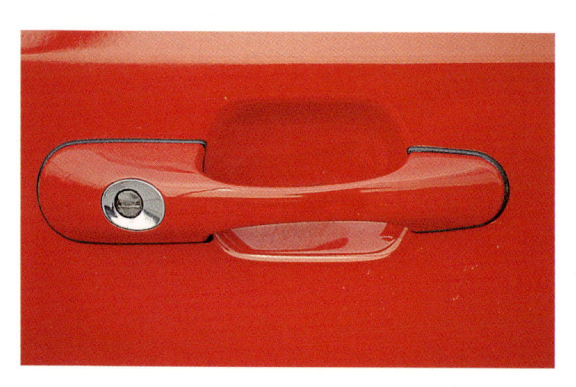

The old VW door handles were pensioned off on the 968, to be replaced by these colour-coded versions. Again there are no straight lines in the design.

The 'italic' style script used on the 911 became standardised across the range on the 968, replacing the block lettering of the 944 family. This US-specification model has the mandatory central brake light.

Out with Dumbo, in with Mickey Mouse! The old 'elephant ear' external mirrors were replaced by these more stylish items, still with electric operation and heating.

968

The interior of the 968 was not greatly changed and remained pure Porsche. Electrically adjustable seats were extremely comfortable for long journeys. This car also has a passenger airbag fitted above the glove compartment. Door trims had a revised design, including an extra speaker and a different side pocket.

Dashboard & Instruments

The dashboard was largely unchanged from the S2, except that there was a 300kph (or 180mph) speedometer. The read-out on the digital display was now for outside ambient air temperature rather than time, there now being a new analogue clock in the centre of the dash. The washer bottle acquired a warning level lamp.

Luggage Compartment

There were no major changes from the outgoing 944S2.

Engine

The 968 brought some exciting developments to the existing 3-litre 944S2 engine.

Porsche, ever conscious that its four cylinders were at least two short of what most sports car drivers considered a complete engine, had made much of the fact the performance of its 'four' was much better than most 'sixes' or 'eights'. Now, with the new two-into-one, larger-bore exhaust system, the engine developed a deeply sonorous twin-cam roar on full throttle to herald its impressive performance. Figures of 240bhp at 6200rpm and maximum torque of 305Nm (225lb ft) at 4100rpm were the best of any normally breathing 3-litre engine of the time. Building on the inherent smoothness of the S2's unit, the goals were now improving specific fuel consumption and

Porsche 924/944/968

Airy and purposeful interior of the Cabriolet, here finished in attractive Classic Grey and fitted with Tiptronic transmission. Driver's view shows new 180mph speedometer and Tiptronic gear selection indicator in the lower part of the rev counter. The steering wheel has the standard bulky airbag installation.

This 968 has a cross-bar to fit racing harness shoulder straps. The split rear seats meant the end of the very practical roller blind luggage area cover.

flattening the torque curve by increasing the pulling power at lower revs, thereby making the engine less 'peaky' when accelerated in one gear. Behind the performance goals stood the requirement for all markets to make the exhaust emissions as clean as possible.

With each cylinder already having a volume of 750cc, increasing the overall capacity was not considered an option. Instead there was a whole series of improvements to the engine management system, intake manifold, camshaft timing, valves and exhaust system, along with work on reducing component weights.

968

Engine bay of the 968 shows off the significant amount of decorative moulding. At the front there is even a container to hold a pair of disposable gloves! Compare this view with that of the Club Sport engine on page 113.

The engine management system controlled the solenoid actuator for the VarioCam variable camshaft timing. The mechanism could alter the timing difference (between the intake and exhaust camshafts) by up to 15°.

To achieve a higher specific power output, the engine was developed to run up to a higher maximum of 6200rpm, compared to the S2's old 5800rpm. The pistons and rods were forged and significantly lightened, with the undersides of the pistons cooled by an oil spray similar to that used on the 911. This weight reduction in turn moved the crankshaft's 'fourth-order' natural frequency of vibration outside the engine's normal rev range, allowing the heavy crankshaft torsional vibration damper, weighing 2.5kg (5.5lb), to be discarded. Compression ratio edged up from 10.9:1 to 11.0:1.

The most interesting development of all was the new variable cam timing mechanism, named 'VarioCam' by Porsche's marketing organisation. VarioCam addressed the requirement for additional mid and low-end torque, while at the same time not compromising the improved high-end maximum power.

The inlet camshaft was driven from the exhaust camshaft by a centre-mounted Simplex chain, tensioned and damped by a spring-loaded plunger mechanism located between the two camshafts. If the plunger mechanism was caused to expand, increased pressure from it made the chain loop longer on one side than on the other, causing slight rotation of the intake camshaft relative to the exhaust camshaft. A maximum timing difference of 15° was so provided when the spring-loaded plunger moved 5.8mm (0.25in). The plunger was actuated by a solenoid, the position of which was controlled by the engine management system according to engine revolutions, load and engine oil temperature.

When current was applied to the solenoid at low engine revolutions (but not below 1500rpm and at a defined engine loading), the plunger filled with engine oil and moved the chain ramp upwards to advance the intake camshaft timing from 'late' to 'early'. The solenoid only reverted the tensioner to conventional form above 5500rpm. Early opening of the intake valves

Porsche 924/944/968

reduced flow reversal of fresh charge back into the intake manifold, so improving volumetric efficiency and torque. When the intake timing was retarded above 5500rpm, the pulsed or supercharged intake effect described later was used to get the intake charge into the combustion chamber to increase power output.

By designing the camshaft profiles for maximum power, and then being able to advance the timing when revolutions fell below a set point, low-range torque could be maintained and exhaust emissions optimised in both high and low ranges. This ability to control the valve overlap in conditions of low revs and high load was a significant contributor to the 60% reduced NOx and 75% reduced HC emissions, as it permitted a form of exhaust gas recirculation through which unburned gases were partially returned to the intake tract.

The Bosch fuel injection was developed beyond the basic L-Jetronic equipment of the 944 to become a fully integrated unit with the Motronic M2.10.1 engine management system. Air/fuel mixture control was achieved by purely electronic inputs to the fuel metering unit, rather than part-electronic and part-mechanical as before. The first stage of this process exactly measured the amount of air entering the engine, allowing precise metering of fuel quantity by a hot-wire anemometer in the inlet ducting which replaced the mechanical air flap previously used on all Bosch injection systems. In addition, the wire method produced less pressure loss than the sensor flap, in turn benefiting engine output.

The control unit within the engine management system balanced a number of inputs, including engine revs and load, against pre-programmed 'maps' of data, and then passed fuel metering data to the sequential fuel injection system.

The intake manifold was designed with a bypass passage which allowed fresh air into the rear of the intake plenum as well as the front. The ducting, carefully shaped for optimum gas flow, was designed to take advantage of the pulsing effect inherent in the air intake flow of any engine, giving an effect similar to low-pressure supercharging, when more air volume is forced into each cylinder than would naturally be the case. This gave significant improvements in volumetric efficiency and torque in the medium to higher rev ranges, with Porsche claiming the brake mean effective pressure – the pressure on the piston at the top of its combustion stroke – was increased by 0.3 bar (4psi) due to the resonance effect.

The Motronic system also offered cylinder-specific anti-knock control and progressive protection against engine over-revving, achieved by shifting the ignition timing. Other 'invisible' functions included adaptive sensing of idle speed and on-board diagnosis with fault memory. The system was fail-safe if the hot wire was damaged, thanks to a 'limp home' facility. There was an interface to the optional electronically-controlled Tiptronic transmission.

From the start of the 1993 model year, cars leaving the factory were also filled with Shell TMO SAE 5W40 synthetic oil.

Transmission

Buyers could choose a six-speed manual or the optional four-speed Tiptronic automatic, a brand new four-speed transmission jointly developed by Getrag and Audi which was noticeable for its light and positive feel.

The manual's long-legged sixth gear allowed the 968 to cruise at 100mph at just over 4000rpm, while on winding roads drivers could use the approximate 500rpm between ratios to keep the engine turning at 4000-5000rpm. This gearbox was a dream.

The same pursuit of driving pleasure could also be obtained from the Tiptronic transmission. The market wanted a good automatic from Porsche, with straightforward two-pedal driving and no shifting of a gear lever required. Tiptronic provided that combination with a level of intelligent functionality which coped with the majority of driving situations. The transmission was derived from the ZF automatic 4HP 18FL unit, subsequently developed by Porsche.

In normal use the Tiptronic lever function operates like a conventional automatic, with P-R-N-D-3-2-1 positions, but with swifter changes. The factory claimed that the reduced number of gears was compensated for by there being no break

Tiptronic shift lever in the fun position! This is 'manual' mode, for sequential gear changing up and down. Other switches, from left, are for the central door lock, external mirror adjustment, cabriolet top and headlamp beam height adjustment. The analogue clock was a new feature on the 968, while the stereo is an aftermarket fitting.

in power delivery during gear changing and, of course, no clutch pedal movement.

On the Tiptronic the conventional hydraulic control of the gear changes was supplemented by a dedicated electronic control unit. Developed by Bosch and programmed with its own operational 'maps', this used 55 inputs and outputs from system components and the Motronic engine management system. To expand the flexibility of a conventional automatic to appeal to the sports car driver, there were five shift maps, starting with economy mode and going up to sport mode. The driver did not even have to worry about selecting the right one of these, as the system monitored accelerator position, lateral acceleration and other driving parameters to sense both driving conditions and style. Fast throttle use and lift were detected as more enthusiastic driving, resulting in late up or down changing, while high lateral accelerations during high-speed cornering were also detected, the control unit then preventing unwanted changes up or down. In contrast, slow accelerator movement resulted in early changes and therefore better fuel consumption.

Meanwhile the slip in the torque converter, always a problem with automatics which results in loss of traction, was minimised, particularly in third and fourth gears. This reduced the kick effect typical of automatic transmissions.

Yet, even with all this, Porsche still wanted to retain the spirit of its defunct Sportomatic transmission – a semi-automatic gear lever offered on the 911 in the '70s – by offering a clutchless manual change. The Tiptronic change gate therefore had two slots – one for full automatic and one for clutchless manual. The lever could be moved across the gate to change from one to the other at any time.

The manual change was similar to a motorcycle's sequential movements – forwards to change up, backwards to change down, with the driver making the choice of when to change, subject only to the maximum revs available in each gear. Not only was this fun, but the protective electronics proved their value when drivers accidentally tried to change up by pulling the lever backwards – easily done if you were not used to this pattern.

For the US market, Tiptronic was also equipped with two other important safety features. 'Keylock' ensured that the ignition key could only be inserted or withdrawn when the shift lever was in the 'P' position, while 'Shiftlock' prevented the driver selecting 'D' or 'R' from park or neutral unless the foot brake was depressed. A useful feature was also that the shiftlock could be released by the brake pedal after the gear lever was moved, rather than having to move it back to 'P' and then press the pedal.

The manual transmission in turn was enhanced by the addition of the option of a new Torsen limited slip differential. The Torsen (TORque SENsing) diff improved tractability on poor surfaces and in tight cornering through improved sensing of changes of grip between the rear wheels and the surface. The diff did not generate a locking value across special friction surfaces, but instead worked from the friction produced by the tooth surfaces of both helical spur gears meshing with six worm gears when different wheel torques or speeds were present on either side of the car.

Electrical Equipment

The headlamps were a significant improvement over those of the 944, featuring a reflector system with variable focus to improve light dissipation. The integrated driving and fog lights had better focused reflectors which improved roadside illumination by up to 30% compared to the S2's units, as well as reducing dazzle in fog.

The car came with a six-speaker sound system, but ex-factory radio equipment was an extra cost option. On offer was the Grundig 5500 RDS or the excellent Blaupunkt Symphony RDS (for Europe and Rest of the World markets) or CR1 (for the US) digital display radio/cassettes with anti-theft coding. There were also three CD changer options.

For 1993, the standard speaker system was improved and key card security provided for the RDS-equipped Blaupunkt London RDM42 CD radio/cassette. For 1994, there was the Bremen RCM43 with key card, while domestic German cars were offered the option of a D-net compatible Motorola cellular telephone installation and the sound system upgraded further still by a new amplifier and front speakers.

Suspension & Steering

The suspension was identical in layout to that of the 944S2. Ride quality mid-way between the S2 and Turbo was provided by MacPherson struts and cast alloy lower control arms at the front and semi-trailing arms and torsion bars at the rear. The standard anti-roll bar package comprised 26.8mm (1.06in) front and 16mm (0.63in) rear bars, but stiffer 30mm (1.18in) front and 20mm (0.79in) rear bars were optional. The rack and pinion steering was power assisted, as before, with 3.25 turns lock to lock.

Although the coupé always provided a firmer ride than the Cabriolet, this contrast is only noticeable at handling extremes and should not unduly influence a buying decision. The Sports suspension package, which came with wider 17in

Porsche 924/944/968

wheels, included bigger brakes, adjustable shock absorbers (Koni instead of the standard Sachs), lower ride height and the stiffer anti-roll bars.

Brakes

As with the 944S2, the ventilated discs had four-piston fixed calipers with Bosch ABS as standard. The front discs were 297mm (11.7in) diameter and the rears 300mm (11.8in).

For 1993, DOT 4-200, a brake fluid with a higher boiling point, was specified, with the suggested replacement interval increased from two to three years. From launch the Sports package included larger front discs of 304mm (11.97in) diameter. For 1994, a new 'special chassis' package featured cross-drilled discs.

Wheels & Tyres

The 7J × 16in and 8J × 16in alloy wheels were styled on the racing Turbo/Carrera Cup five-spoke design. Standard tyres were 205/55ZR16 front and 225/50ZR16 rear, which helped lower the aspect of the car to the ground. The factory-recommended tyre choices were Bridgestone Expedia S-01 N2, Continental CZ 91 NO, Toyo Proxes F1 S NO, Goodyear Eagle GSD N2 and Yokohama A008P NO.

As part of what was termed the Sport package, 7.5J × 17in and 9J × 17in alloys with 225/45ZR17 and 255/40ZR17 tyres were optional (option M398). These improved the car's cornering power significantly, but only at the cost of increased interior noise and a harsher ride. For these wheels,

The pressure-cast Cup Design wheel (near right) was the discovery of the '90s in its ability to enhance the looks of any Porsche, filling the void left by the fashionable Fuchs forged alloys of the '80s. Variations on the theme show the narrow-spoke version (centre right) introduced for 1994 and the more rounded design (far right) that came with the Sport package.

968

Michelin MXX3 and Pirelli P70-ZNO joined the list of recommended tyres and Toyo was dropped.

For 1994, the appearance of the wheels was revised, with narrower, more rounded section spokes and new wheel-locking nuts.

968 Club Sport (1993)

Despite the 968's unmatched refinement, it was clear immediately after launch in August 1991 that the car was not going to light up the interest of a market in recession. In the US, a 968 could be bought for under $40,000 – a price designed to win new customers. But in Europe, where the price was equivalent to $54,000, potential buyers were not impressed and sales as a result were extremely disappointing.

Drastic action was needed to save the model and therefore the 968 Club Sport, costing 17% less than the standard car, was launched in October 1992, and then in right-hand drive form in the spring of 1993.

The Club Sport was very basic, in the mould of the 911RS, and only offered in five striking colours – Black, Maritime Blue, Guards Red, Grand Prix White and Speed Yellow. The spoiler was colour-coded to the body and a large 'Club Sport' script in red, white or black was plastered along each side – unless a customer chose the no-cost option of deleting it.

The Club Sport also had the standard car's

The Club Sport stormed onto the 968 scene in 1993 to boost flagging sales in Europe. Stripped of most luxury equipment and denied the mainstream model's inherent practicality, it was a 968 for the race track or Sunday mornings. The stance of the Club Sport is noticeably lower than that of the standard 968, although the difference in ride height is only 20mm.

111

Porsche 924/944/968

optional 7.5J × 17in and 9J × 17in wheels, with 225/45ZR17 and 255/40ZR17 tyres, either colour-coded to the body or otherwise silver. The 3-litre engine was unchanged, as were the outstanding brakes, with ABS retained. The suspension was lowered 20mm (0.8in) and the car had firmer handling than the standard 968, yet a ride which was still acceptable for everyday motoring.

The external mirrors were manually adjusted and there was no rear wiper, heated headlamp washers or roof aerial (unless a radio was fitted as an option). Inside there were two lightweight, narrow Recaro seats, their backs colour-coded to the body. Each Recaro weighed 13.6kg (30lb), compared to the electric seat's 22kg (48.5lb). The driver had the chunky 360mm (14in) Sports package steering wheel, with no airbag, while the windows had manual winders and the heater was manually adjusted.

The alarm system was also removed, but what was considered the worst saving of all was the change to a manual release for the rear hatch. This meant the glass tailgate could not be opened from the outside, while the toggle release was impossibly located on the left rear wheel arch behind the driver's seat.

As the Club Sport was a strict two-seater, a glass-fibre board covered the rear seat buckets and there was no luggage cover or luggage supports. The whole of the rear area was covered in lightweight material instead of the usual fitted upholstery, while a Blaupunkt Paris radio/cassette with RDS was only offered as an option (although it was standard in some markets), replacing the normal 968's Symphony unit.

In the engine compartment, the styling mouldings were discarded, there was only one radiator cooling fan and noise insulation was removed. Porsche also claimed the wiring loom had been reduced, incidentally making electrical upgrading difficult. The alternator was reduced to a 90 amp unit and the battery to 36Ah from 63Ah. If air conditioning was specified (option M573) the larger battery, alternator and normal double cooling fan arrangement was refitted.

Porsche claimed a weight reduction over the standard car of 50kg (110lb), down to 1320kg (2910lb), although in markets like the UK, where a number of factory options were fitted as standard to the regular 968, the difference was more like 100kg (220lb).

Among the list of options, which allowed the car to be re-equipped upwards to virtually the specification of a standard 968, was the Sports package, coded M030. This included stiffer springs, shock absorbers and anti-roll bars, drilled brake discs, with the fronts enlarged to 304mm (11.97in) diameter, and a 40% Torsen limited slip differential; the diff (option M220) could be ordered separately. There was also a 'Security' package including central locking, an alarm system and locking wheel nuts. Meanwhile the 'Comfort' package included electric external door mirrors and electric windows.

The Club Sport's aggressive look was strengthened by colour-coded wheels, although silver was a no-cost option – as was deletion of the enormous side decals.

968

The engine bay of the Club Sport has all non-essential plastic trim removed. In the right foreground is the brake fluid reservoir on the vacuum servo. The yellow-capped oil filler is immediately behind.

The interior of the 968 Club Sport was basic. This view shows the backs of the glass-fibre seats, which saved some 8kg compared to the standard electric versions. The thick-rim 380mm non-airbag steering wheel was a delight to handle.

Porsche 924/944/968

This time the reaction of the press was markedly different. The car was received with open arms, with *Autocar & Motor* awarding it its prestigious 'Car of the Year' award in 1993 for its performance, handling and driving experience. Porsche, readers were told, had discovered real sports cars!

Meanwhile the company preferred to remain conservative about any improvement in straight-line performance. Partly to ensure that this limited-production car did not need new type approval, 0-100kph (62.5mph) and top speed remained officially unchanged at 6.5sec and 156mph (251kph). However, with the full weight reduction, the acceleration time was actually approximately 0.2sec less to 100kph, while the lower ride height permitted 160mph (257kph).

The Club Sport helped Porsche sell 968s in Europe, but it was never going to be the solution in a market which was still slow – although it might have nurtured huge enthusiasm among club racers and those using race circuits for enjoyable driving days or for driver education.

968 Turbo S (1993)

Back in 1980, the 924 Carrera GTR had demonstrated that the base model's well-balanced chassis could soak up more power without drama, as had the later 944 Turbo S. So when the 944 became the 968, everyone wondered whether a yet more powerful model would be developed. But the reality had been that the 944 Turbo had not sold as well as had been expected, meaning that a mainstream 968 Turbo was never really going to be economically feasible.

But after the mediocre reception accorded to the car at launch, Porsche decided it was important that the character of the new model was injected with more fizz, so the development of a Turbo S was run alongside that of the Club Sport. Although the Turbo S could be road registered, as with the 924 Carrera GTS 12 years previously, customers could only buy the car direct from the Customer Sport Department in Weissach.

The specification baseline was the Club Sport, with the already stripped-out body receiving detail improvements to improve airflow both into and over the car. The front air dam was given a splitter at its lower edge, and the opening above the number plate enlarged, while at the rear a small aerofoil spoiler on the hatch could be adjusted through a range of 10°. Two small NACA inlets on each side of the bonnet provided cooling air to the engine compartment, while, to save weight, PVC underbody protection was omitted. Inside, the specification was similar to the Club Sport with, in essence, everything taken out. Meanwhile, of course, owners had the extensive options list which enabled them to put most of the equipment back again if they so desired.

The chassis itself was lowered 20mm (0.8in) with harder springs and shock absorbers, while the car crouched on massive 235/40ZR18 front and 265/35ZR18 rear tyres. The 8J and 10J wheels were the extremely attractive Cup Design Speedline split-rim cast alloys, which made proud owners, and non-owners, drool, accompanied by the 911 Turbo's red-painted four-piston calipers over cross-drilled ventilated discs. ABS was standard equipment.

Instead of the VarioCam four-valve cylinder head, the 3-litre block was fitted with the 944 Turbo's proven eight-valve head and the same compression ratio of 8.0:1 (M44/60).

The 968 Turbo S delivered 305bhp at 5400rpm, with huge maximum torque of 500Nm (369lb ft) at 3000rpm. Boost was 1.0 bar from the water-cooled KKK turbocharger. Despite the very high torque figure, the new six-speed 968 gearbox was still used, giving the car awesome overtaking potential on the way up to its maximum speed of 175mph. Fourth and fifth, as well as the crown wheel and pinion, received longer ratios, but otherwise the gearbox was unchanged, although the clutch was strengthened to cope with the extra torque. A limited slip differential with a 75% locking factor was fitted as standard equipment. In all, the Turbo S weighed 70kg (154lb) less than the regular car.

'Club Sport' logos complemented the five striking body colours offered. This is Speed Yellow – a Weissach favourite for hotter Porsches during the mid-'90s. The Club Sport was strictly a two-seater – and practicality as a luggage carrier was diminished. The hatch release was blanked off from the outside and acrobatics were needed to get at the handle behind the driver's seat.

The 968 Sport was a UK-only model which combined the suspension of the Club Sport with all the useful accessories of the regular car. It was what the 968 should have been from the beginning and sold relatively well.

The even quicker 968 Turbo RS was developed for customers to compete in the German ADAC GT Cup. Like the 924 Carrera GTR, it is not strictly relevant in this book because it is not a road car, yet it is impossible to ignore the fact that its output was 337bhp, with a Le Mans specification version with 350bhp even on offer. The 10J × 18in and 11J × 18in rims were equipped with 265/630 and 305/650 racing slicks, and weight was adjusted upwards to 1350kg (2976lb) to equate with the maximum 4kg/bhp power/weight ratio allowed for the ADAC class.

968 Sport (1994-95)

While many markets used price discounting and factory options to increase the attractiveness of the standard car, British customers, aware how much it cost across the Atlantic, demanded a model which had the equipment level of the standard 968, yet at the price of the Club Sport. This sounded like a tall order from a business viewpoint for Porsche, but in these desperate times it was answered by the 968 Sport, which was released in January 1994 as a UK-only model.

The Sport kept the basic dynamics of the Club Sport, including the lowered ride height, but added a few luxuries for an extra £3000 on the Club Sport's 1994 price of £29,975, including central locking, alarm, electric windows and mirrors, rear seats and the welcome return of the electric tailgate release. The factory's claimed performance was identical to the Club Sport, with weight increased by only 30kg (66lb).

In Britain, there seemed little reason to consider spending another £5500 on top of that to acquire a standard 968, so not surprisingly during 1994-5 a total of 306 Sports were sold, compared to only 40 coupés and 71 Cabriolets.

The ultimate development of the mainstream production four-cylinder Porsche, the Sport sparkled in every area the enthusiast could wish, above all by not sacrificing the enduring feature of complete practicality for everyday use.

Porsche 924/944/968

Data Section

IDENTIFICATION DATA

Year	Model	Market	Chassis numbers	Engine type	Gearbox type
1992	968	RoW	WPOZZZ96ZNN800001-0006	M44/43	G44/00
	968	US	WPOAA296_NN820001-0004	M44/43	G44/00
	968 Cab	RoW	WPOZZZ96ZNN830001-0003	M44/43	G44/00
	968 Cab	US	WPOCA296_NN840001-0002	M44/43	G44/00
	968	RoW	WPOZZZ96ZNS800001-2541	M44/43	G44/00
	968	US	WPOAA296_NS820001-0709	M44/43	G44/00
	968 Cab	RoW	WPOZZZ96ZNS830001-1366	M44/43	G44/00
	968 Cab	US	WPOCA296_NS840001-0727	M44/43	G44/00
1993	968	RoW	WPOZZZ96ZPS800001-1203	M44/43	G44/00
	968 CS	RoW	WPOZZZ96ZPS815001-5856	M44/43	G44/00
	968	US/Canada	WPOAA296_PS820001-0668	M44/43	G44/00
	968 Cab	RoW	WPOZZZ96ZPS830001-0631	M44/43	G44/00
	968 Cab	US/Canada	WPOCA296_PS840001-0414	M44/43	G44/00
	968 Turbo S	RoW	WPOZZZ96ZPS890061-0071	M44/60	G44/01
1994	968	RoW	WPOZZZ96ZRS800001-0298	M44/43	G44/00
	968 CS	RoW	WPOZZZ96ZRS815001-5536	M44/43	G44/00
	968	US/Canada	WPOAA296_RS820001-0778	M44/43	G44/00
	968 Cab	RoW	WPOZZZ96ZRS830001-0128	M44/43	G44/00
	968 Cab	US/Canada	WPOCA296_RS840001-0741	M44/43	G44/00
	968 Turbo S	RoW	WPOZZZ96ZRS890061-0063	M44/60	G44/01
1995	968 Cab	RoW	WPOZZZ96ZSS830061-0061	M44/43	G44/00
	968 CS	RoW	WPOZZZ96ZSS815001-5531	M44/43	G44/00
	968	US/Canada	WPOAA296_SS820001-0258	M44/43	G44/00
	968 Cab	US/Canada	WPOCA296_SS840001-0366	M44/43	G44/00

Note
The records show that just three Turbo RS models were built in 1993 (chassis numbers 96PS896061-63).

PRODUCTION DATA

Year	Model/market	Max power (bhp@rpm)	Max torque (Nm@rpm)	Compression ratio	Weight (kg)	Number built
1992	968 RoW	240@6200	305@4100	11.0:1	1370	2547
	968 US	240@6200	305@4100	11.0:1	1370	713
	968 Cab RoW	240@6200	305@4100	11.0:1	1370	1366
	968 Cab US	240@6200	305@4100	11.0:1	1370	727
1993	968 RoW	240@6200	305@4100	11.0:1	1370	1203
	968 CS RoW	240@6200	305@4100	11.0:1	1320	856
	968 US/Canada	240@6200	305@4100	11.0:1	1370	668
	968 Cab RoW	240@6200	305@4100	11.0:1	1370	631
	968 Cab US/Canada	240@6200	305@4100	11.0:1	1370	414
	968 Turbo S RoW	305@5400	500@3000	8.0:1	1300	11
1994	968 RoW	240@6200	305@4100	11.0:1	1370	298
	968 CS RoW	240@6200	305@4100	11.0:1	1320	536
	968 US/Canada	240@6200	305@4100	11.0:1	1370	778
	968 Cab RoW	240@6200	305@4100	11.0:1	1370	128
	968 Cab US/Can	240@6200	305@4100	11.0:1	1370	741
	968 Turbo S RoW	305@5400	500@3000	8.0:1	1300	3
1995	968 Cab RoW	240@6200	305@4100	11.0:1	1370	1
	968 CS RoW	240@6200	305@4100	11.0:1	1320	531
	968 US/Canada	240@6200	305@4100	11.0:1	1370	258
	968 Cab US/Canada	240@6200	305@4100	11.0:1	1370	366
Total						12,776

Note
1992 models include cars built at the end of the 1991 model year (see table above).

IDENTIFICATION

Chassis numbers The first 60 numbers in each model year were reserved for internal factory use, although they were not necessarily used. Therefore, the end production serial number may not be the number which had been built. This rule did not apply to the 968 Turbo S and RS models. Factory records show that no 968 coupés were built for RoW markets for the 1995 model year.

From May 1993 all models for US/Canada were sold as R programme (ie, the upcoming model year) vehicles. These are identified by the M718 option code. The same happened in 1994 – namely the M718 option was given an S model serial number.

Engine numbers Manual transmission engines were M44/43, with serial numbers beginning 42N (1992 model year), 42P (1993 model year), 42R (1994 model year) or 42S (1995 model year); these prefixes were followed by a five-digit serial number. Turbo S and RS engines were M44/60. Generally the first 500 numbers in the serial were reserved for factory use.

Gearbox numbers Serial number series for manual gearbox was G44001000001-5317 in 1992. With limited slip differential (M220), sixth digit changed to 2 (ie, G44002002001-4799 in 1992). These numbers exclude numbers 1-2000 reserved for internal factory use, although again not necessarily used. Tiptronic engines were M44/44 with the first digit of the five-digit serial number starting 5; Tiptronic was designated A44/00.

Six-speed manual transmission ratios: 1st, 3.182; 2nd, 2.000; 3rd, 1.435; 4th, 1.111; 5th, 0.912; 6th, 0.778; reverse, 3.455; axle ratio, 3.778. Optional Tiptronic automatic transmission ratios: 1st, 2.579; 2nd, 1.407; 3rd, 1.000; 4th, 0.742; reverse, 2.882; axle ratio, 3.250.

With all the appearance trim in the engine compartment, it needed a few labels to remind that an engine did exist under the plastic.

968

Key Facts
Fuel European version: 98 RON unleaded.
Dimensions Length, 4320mm; width, 1735mm; height, 1275mm; front track, 1472mm; rear track, 1450mm; wheelbase, 2400mm; ground clearance (empty), 125mm; turning circle, 10.75m; kerb weight, 1370kg (coupé), 1440kg (Cabrio), 1430kg (coupé Tiptronic), 1500kg (Cabrio Tiptronic).
Capacities Fuel tank, 74 litres; engine oil, 7 litres; coolant, 8 litres; transmission (manual), 3.5 litres; screen/headlamp wash, 3 litres.

Options
1992
M058 Energy absorbing bumpers, front and rear; M139 Heated front seat (passenger); M220 Limited slip differential; M288 Headlamp washers; M340 Heated front seat (driver); M383 Sport seat with electric height adjustment (passenger); M387 Sport seat with electric height adjustment (driver); M490 Hi Fi sound system with ten loudspeakers (eight for Cabrio) plus amplifier; M498 Deletion of model designation; M513 Lumbar support, driver's seat (only with M438); M573 Automatic temperature control with air conditioning (standard in US); M463 Windscreen clear, not tinted; M526 Cloth door panels; M586 Lumbar support, driver's seat (only with M437); M454 Automatic speed control with resume feature (standard in US); M595 Colour-coded rear spoiler; M650 removable sunroof panel (standard in most markets); M327 Blaupunkt Symphony RDS stereo radio/cassette; Porsche CD2 digital display AM/FM CD stereo; M690 CD10 Radio/CD player plus anti-theft; RA01/03 Remote-controlled ultrasonic alarm (for models with and without sunroof); RR04 Grundig 5500 radio/cassette plus RDS; RR05 Sony Disc Jockey Radio/Cassette/CD; Porsche CR1 digital display stereo/cassette (standard in US); M692 CD changer; Sport chassis (coupé only); 17in five-spoke Turbo design cast alloy wheels; Leather seats; Partial leather front seats; Leather interior trim; All cloth front and rear seats; Metallic paint; Paint and chrome protection deletion.

1993
M030 Sport Chassis; M693 CD-Radio Blaupunkt London RDM42 plus code card (only with M490); Right-hand drive passenger airbag; M567 Colour-coded rear spoiler; Portable cellular telephone with integrated console and hands-free operation; Piping to front seats.

1994
P31 Sport chassis M398 plus stiffer springs, shock absorbers externally adjustable, adjustable spring plates, adjustable stabiliser, stronger suspension bushes at front; M334 Blaupunkt Bremen RCM42 plus code card; M398 New Cup wheels – 7.5J x 17in with 225/45ZR17 tyres front, 9J x 17in with 255/40ZR17 tyres rear; M418 Protective side mouldings; M437/8 Driver/passenger front seats with electric height, length and lumbar adjustment.

Colours & Interiors
1992 (Porsche Cars Great Britain chart)
Solid colours A1 Black, F2 Maritime Blue, G1 Guards Red, G4 Rubystone Red, M1 Signal Green, N4 Mint Green, P5 Grand Prix White, 98 Colour to sample (cost option).
Metallic colours A8 Polar Silver met, F4 Horizon Blue met, N7 Amazon Green met, N9 Oak Green met, F6 Cobalt Blue met, F9 Amethyst met, G7 Coral Red met, Q9 Slate Grey met, Z8 Black met, 99 Metallic colour to sample (cost option).
Special metallic colours 50 Satin Black met, 51 Venetian Blue met, 52 Cassis Red met, 53 Granite Green met, 54 Lagoon Green met, 55 Zermatt Silver met, 56 Marine Blue met, 57 Violet Blue met, 58 Tahoe Blue met, 59 Turquoise met.
Upholstery Leatherette: Magenta, Cobalt Blue, Marine Blue, Classic Grey, Black. Combination colours: Light Grey/Magenta, Light Grey/Cobalt Blue, Light Grey/Classic Grey, Light Grey/Black, Cashmere Beige/Black. First colour is used for carpet, main seat material and door inlay. Second colour is used for dash, knee bar, roof liner, window pillars, sun visors, window sills front/rear, rear backwall section and steering wheel (leather). Leather: Magenta, Cobalt Blue, Marine Blue, Classic Grey, Black. Combination colours: Light Grey/Magenta, Light Grey/Cobalt Blue, Light Grey/Classic Grey, Light Grey/Black, Cashmere Beige/ Black. Special leather: Matador Red, Carrera Grey, Sherwood Green (plus carpets in same colours) or own order sample (cost option). 'Porsche' cloth: Magenta, Black, Cobalt Blue, Marine Blue, Classic Grey.
Carpet Magenta, Black, Cobalt Blue, Marine Blue, Classic Grey.
Cabrio soft-top Magenta, Black, Cobalt Blue, Marine Blue, Classic Grey.

1993 (Porsche Cars Great Britain chart)
Solid colours A1 Black, F2 Maritime Blue, G1 Guards Red, M1 Signal Green, N4 Mint Green, X4 Speed Yellow, P5 Grand Prix White, 98 Colour to sample (cost option).
Metallic colours A8 Polar Silver met, F4 Horizon Blue met, N7 Amazon Green met, N9 Oak Green met, F6 Cobalt Blue met, F8 Midnight Blue met, F9 Amethyst met, A7 Rhodamine Red met, Q9 Slate Grey met, Z8 Black met, B5 Wimbledon Green met, 57 Violet Blue met, 99 Metallic colour to sample (cost option).
Upholstery Leatherette: Magenta, Cobalt Blue, Marine Blue, Classic Grey, Black. Combination colours: Light Grey/Magenta, Light Grey/Cobalt Blue, Light Grey/Classic Grey, Light Grey/Black, Cashmere Beige/ Black. Leather: Magenta, Cobalt Blue, Marine Blue, Classic Grey, Black. Combination colours: Light Grey/Magenta, Light Grey/Cobalt Blue, Light Grey/Classic Grey, Light Grey/Black, Cashmere Beige/Black. Special leather: Matador Red, Carrera Grey, Sherwood Green, Rhodamine Red, Wimbledon Green, Firenze Grey (plus carpets in same colours) or own order sample (cost option). 'Porsche' cloth: Magenta, Black, Cobalt Blue, Marine Blue, Classic Grey.
Carpet Magenta, Black, Cobalt Blue, Marine Blue, Classic Grey.
Cabrio soft-top Magenta, Black, Cobalt Blue, Dark Blue, Classic Grey.

1994 (Porsche Cars Great Britain chart)
Solid colours A1 Black, S8 Riviera Blue, G1 Guards Red, T3 Amaranth Violet, X4 Speed Yellow, P5 Grand Prix White, or colour to sample (cost option).
Metallic colours A8 Polar Silver met, D3 Iris Blue met, K6 Aventura Green met, F8 Midnight Blue met, Q9 Slate Grey met, Z8 Black met, 99 Metallic colour to sample (cost option).
Upholstery Leatherette: Midnight Blue, Classic Grey, Chestnut Brown, Black. Leather: Classic Grey, Chestnut Brown, Midnight Blue, Black. Special leather: Flamenco Red, Cedar Green or own order sample. 'Porsche' cloth: Black, Midnight Blue, Chestnut Brown, Classic Grey.
Carpet Midnight Blue, Chestnut Brown, Classic Grey, Black, Marble Grey, Cashmere Beige.
Cabrio soft-top Midnight Blue, Chestnut Brown, Classic Grey, Black, Marble Grey.

1995 (charts WVK 127 420 95, VM 8/94)
Solid colours A1 Black, G1 Guards Red, P5 Grand Prix White, S8 Riviera Blue, T3 Amaranth Violet, X4 Speed Yellow, or colour to sample (cost option).
Metallic colours A8 Polar Silver met, D3 Iris Blue met, F8 Midnight Blue met, K6 Aventura Green met, Q9 Slate Grey met, Z8 Black met, 99 Metallic colour to sample (cost option).
Upholstery Leatherette: Black, Classic Grey, Chestnut, Cedar Green, Flamenco Red, Midnight Blue. Leather: Marble Grey, Midnight Blue, Chestnut, Classic Grey, Cashmere Beige, Black. Special leather: Provence Blue, Cedar Green, Flamenco Red. 'Porsche' cloth: Black, Classic Grey, Cashmere Beige, Marble Grey, Midnight Blue, Chestnut.
Carpet Black, Classic Grey, Cashmere Beige, Marble Grey, Midnight Blue, Chestnut, Provence Blue, Cedar Green, Flamenco Red.
Cabrio soft-top Black, Classic Grey, Marble Grey, Dark Blue, Chestnut.

Three options shown here are the ultrasonic alarm system indicated by side window graphics (left), automatic temperature control with air conditioning (above) and the colour-coded rear spoiler (right).

Porsche 924/944/968

Buying & Driving

The water-cooled four-cylinder models have introduced Porsche driving to hundreds of thousands of people and, because of their excellent value, are extremely popular in used car markets.

This chapter will consider separately the three main groupings – 924, 944 and 968 – addressing desirability, performance and, most importantly, everyday practicality. As with any model, all have problems to look out for down the years, but you should ignore sour comments about these cars not being 'real' Porsches. In reality the water-cooled cars have found their rightful place among the best, and driving pleasure and excitement can be found from the earliest 924 onwards. Try one on snow and you will revel in the near-ideal weight distribution and safe handling.

From the first models power output progressively increased, seeking, but never quite finding, the limits of a chassis which was so accomplished that it left rival products trailing. Even though the 924 and 944 were greatly copied by Japanese manufacturers, particularly in the '80s, none had the calibre of the Porsche. Whether your budget runs to a 2-litre 924, or goes all the way to a late model 968, you will not be disappointed.

Driving a Porsche is still the purest form.

Basics

Depreciation on the four-cylinder cars has not been as stable as on the 911s. In the early '90s it was significant, but seemed to stabilise after production of the 968 ceased. As most readers will know, the first three years are the worst for loss of value, so the best value will always be found in cars which are at least that age. You should also not forget that older Porsches hold their value because of their sound build quality, reliability and resistance to corrosion. That said, however, the general rule is still to stretch to the most recent model you can afford so you can benefit from the improvements made virtually every year.

Balancing availability, price and desirability in today's market, the mainstream 944 must come near the top of the pile in the four-cylinder family of Porsches. And what better paintwork for one than Guards Red – the most popular Porsche colour of the '80s. This example has the no-cost option of deleting the model designation on the rear.

Buying & Driving

Be very wary of the indicated mileage on any car, especially if more than five years old. Expect to be shown either a full annual statutory inspection record (if applicable), or regular service bills which show the mileage building up, along with any statutory inspection certificates stating the same. Evidence of regular servicing in the car's records is important, particularly items such as changes to oil, brakes, filters, clutch and so on.

The 944 engines, particularly the pre-1986 models with smaller oil capacity, have aluminium bores and bearing journals which need looking after. Unscrupulous sellers do try to pass on cars which are in imminent need of a new clutch, or worse. Yet, if the initial choice has been careful, operating costs can be very reasonable and certainly no more than a normal family saloon.

Generally spares availability is not a problem for all models and official dealers can sell you most parts you will need, with the possible exception of more obscure items for a 924. But, especially if you have an older model, the first place to look is the advertising in club magazines. The better specialists buy exactly the same parts direct from Porsche's suppliers, yet do not charge nearly as much, although the advice is always to stick to well-known brand names and use official parts whenever the braking system is involved.

Official Centres (dealers) are probably the most expensive places to buy a used car, but then they generally only stock vehicles up to between five to seven years old, with good histories and carrying comprehensive warranties, leaving it for you to decide how much this extra peace of mind is worth.

Buying privately obviously requires more effort and carries more risk, which can range from a subsequently-challenged title of ownership, to a hidden mechanical or electrical disaster waiting to occur as soon as you have parted with your money. Unlike buying from an official dealer, it is for you to satisfy yourself that the car's paperwork is good and the overall condition what you expect.

But buying privately is also cheaper, which is why so many people do it, and there are various checks you can carry out, or have carried out, to validate title. Most specialists will also offer a check-over service, which is money well spent if you do not feel confident enough to assess a car yourself. Remember, too, that any specialist, even though being paid a fee, will only identify what they can see. Never expect any later liability for unforseen problems.

Magazines and newspapers carrying classified adverts for sports cars will probably be well-known to the reader, with the widest selection of Porsches always found in the club magazines or newsletters. Guidelines to prices and availability can be gained from the directories in certain of the monthly magazines and by recording price data over a period of time.

The last word is on insurance. There is more to this than simply shopping around, as the companies that advertise are not always the best. Always ask other Porsche drivers first. You will find most clubs or local regions able to advise on where to find the best deals.

Buying a 924

It is the technology and the level of refinement which separates the early 2-litre 924s from the 2.5-litre and later models. That said, there are still some exceptional buys hidden within early cars.

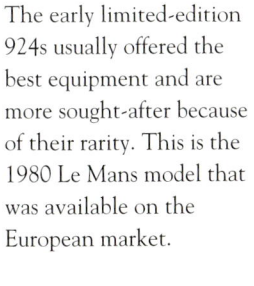

The early limited-edition 924s usually offered the best equipment and are more sought-after because of their rarity. This is the 1980 Le Mans model that was available on the European market.

119

Porsche 924/944/968

The most desirable early 924s are those post-1980 models equipped with the Audi five-speed gearbox, rather than the earlier Getrag five-speed 'box with dog-leg first gear. Pre-1980 cars are lightweight and therefore good for racing, for which the Getrag 'box is much more suitable than the early four-speed unit.

By 1980, the engineers had fairly well designed out the inherent VW-Audi feel of the first cars and turned the 924 into a fine performer. European cars adopted the breakerless transistor ignition and, for 1981, the well-known hot starting problem was pretty much eliminated. Other notable areas which improved dramatically were noise reduction, particularly with the 1983 and later models, and the interior trim, especially from 1982, when the 924 shared its interior with the 944. The nicest fabrics to modern eyes include the Berber and pinstripe.

The cars had all-over galvanising from the start of the 1978 model year, with the Longlife guarantee extended from six years to seven years for the 1981 model year. Fully galvanised cars do seem to last longer.

The three-speed automatic will appeal to those needing two-pedal driving, but you will find performance noticeably degraded. Among the most desirable limited editions were the 1980 Le Mans model and the US-only 1981 Weissach model. Both benefited from the improved looks of the 924 Turbo's rear spoiler, not seen on the regular cars until the 1983 model year, along with more attractive external and internal trimming. But these days they are very rare.

In sports terms, the only popular choice is the Series 2 924 Turbo – the model fitted with Digital Motor Electronics and identified by the side repeaters on the front wings. Earlier Turbos were less refined, with their turbochargers prone to high rates of wear, while their two-tone paint schemes were very contemporary and today look decidedly dated.

The 924 Carrera GT remains the classic development of the 2-litre era and therefore carries a premium to suit. Yet it is also the Cinderella among Porsches, as prices have fallen at the same rate as the other models, undervaluing it in terms of both performance and rarity. The car fully deserves the award of the prestigious Carrera title – it was the only four-cylinder Porsche to earn this. When buying one of these cars, the biggest factor to consider is originality. Because they are very fast, crash damage is an ever-present concern, although evidence of repair work can usually be identified, as on any 924 or 944, by uneven panel gaps, particularly around the headlamps, bonnet and doors. The polyurethane panels have tended to distort with time, so fasteners become visible.

The very rare 924 Carrera GTS is the ultimate road-going development of the classic four-cylinder Porsches.

Ten Tips when Buying a 924

1 Brakes The ex-VW front discs last an average of 30,000 miles before needing replacement. Visually inspect the front discs for scoring or damage, and if the car pulls to one side beware of seized calipers or badly adjusted rear drums.

2 Steering It is quite easy to distort the pressed metal lower suspension A-arm, for example by contact with a high kerb. Watch for wandering or pulling to one side and check the steering play. Although the column was praised for its double elbow safety design, these joints wear. You should not be able to move the road wheel without the steering wheel moving as well. If there is a problem it could only be worn track rod ends, but a new column will be expensive.

3 Noisy transmission Do not worry about torque tube bearings whirring, as long as this noise goes away when you drop the clutch when stationary. But do look out for worn rear wheel bearings or gearbox constant velocity joints, denoted by an audible whirring or 'chuffing' from the rear when in motion, exclusive of which gear you are in. Avoid early four-speeders, with their imprecise change, and only choose an automatic if you really need to have one. And while the Getrag transmissions are good for fast use, the dog-leg first is a pain around town. You should be able to find an Audi five-speeder, which uses the same transmission as the 944, for virtually the same money.

4 High mileage and noisy engines Unless the engine has had extremely good care through its life, expect cylinder head work around 80,000 to 100,000 miles. Signs of worn valve guides are excessive mechanical noise when the engine is cold. The oil spray tube running above the camshaft can break off due to vibration, usually resulting in very worn and noisy lobes on number four cylinder. The oil spray bar can be seen with the oil filler cap removed, while better Porsche specialists can advise on an easy modification which improves the lubrication to the Turbo's cylinder head.

5 Hot starting problems Fuel vaporisation was the ultimate killer of enjoyment on an early 924. From the start of the 1981 model year the problem was pretty much solved by adopting a stronger fuel pump, non-return valve and other detail tweaks which overcame the problem. The one-way valve, fitted from around 1978, is the main fix. If you are inspecting an earlier model, look for it on the exit from the fuel pump. Otherwise it can always be fitted. A strong battery to drive a recalcitrant starter also helps.

6 Unreliable electrics To the average mechanic the 924 seemed like a test bed for an electrical switch manufacturer. In its time it must have had more relays than any other car, and as age and moisture get to the electrics, so does corrosion. Check that everything works and ask if there have been any specific problems. The most common failures on 924s are due to problems such as the fuel pump relay, cold start relay and electrical accessories. A car which will not start is often the victim of a dead fuel pump relay. Always be wary of lots of 'gaffer' tape under the dash, as this indicates unscheduled repairs.

7 Turbocharger The Series 1 924 Turbos were notable for raw power, huge turbo lag and, unfortunately, turbos which did not last long. Owners were not used to having to look after the stressed bearings and seals, which would burn when the engine was shut down after a hard run or subjected to prolonged stop/start driving, cooking the oil in the red hot turbo. Consequently bearings and seals wore out very quickly and an improved arrangement did not come until the Series 2 924 Turbos. The whole problem was not really solved until the advent of the 944 Turbo, which featured improved oil recirculation after shutdown through better crankcase breathing. Today the advice is to only buy a Series 2 924 with evidence of a new turbo in the last 20,000 miles. Look for regular 3000-mile oil changes (or 6000 miles if synthetic oil is used) and note the way the seller shuts down the engine after the demonstration run. The engine should be left to idle for a minute after stopping, in order to let the oil cool.

8 Dampness inside Check the carpets behind the front seats for wetness, which will tell you the sunroof seal is leaking. Look in the wells behind the rear wheels and check the small neoprene grommets are allowing water to drain away. Wetness here indicates a leaking rear hatch seal. If the area is damp and rusting, forget the car. Under the bonnet check behind the engine firewall, where water can leak and wet the front carpets, usually through cracked sealant. This area is also famous for collecting dead leaves and wet debris, particularly around the battery. Check that the drains into the engine compartment are clear and preventing water from standing.

9 Interior wear and damage Leatherettes and leather wear far better than cloths, but of these the darker greys and blacks show wear less than the lighter colours. On older cars seek out black leatherette or leather with black pinstriping, as everything else tends to look tatty after a time. Search very carefully for rips, cigarette burns and parting of seams, particularly on the seat

Porsche 924/944/968

Smooth, flexible power gives the 3-litre 944S2 superb driveability, but its technical complexity means a cast-iron service history is essential.

Check out gaps around the doors, bonnet and headlamps for evidence of earlier crash damage.

backs and rear headlining, where pushing in unusual loads in order to close the glass can have caused damage. In warmer climates check the dash for cracking.

10 Crash damage The 7mm gaps around the opening panels are archaic in motor industry terms, but provide a very good indicator as to whether the car has been in an accident as it is difficult to line them all up again. Post-1980 cars had the seven-year all-over guarantee against rust perforation, which may not have stopped rust by now, but will certainly have helped. Check too for corrosion where the galvanising layer has been broken, for example by cheap filling after a crash.

Buying a 944

The 944 was an immediate success and there is no shortage of good cars on the market, meaning buyers can afford to be very choosy and only pick a car which exactly meets their requirements. Described very simply, the first 944s had the new 2.5-litre Porsche engine installed into the rolling chassis of the Series 2 924 Turbo, along with a revised bodyshell.

With pre-February 1985 cars, it is the engine which received all the development effort to iron out several bugs. However, it is still important to get these in perspective. The new engine was very good and most owners whose cars were regularly serviced had no problems. Top of the list of problems which did occur, however, were the camshaft timing belt and exhaust side engine mounting, both of which were subjected to intense development and by now should have had attention paid to them, one way or another.

Cars built with the 'oval' dash benefited from the development of the Turbo, receiving many of the more powerful car's improvements. The 2.7-litre 944 was the last of the line and definitely the best of the bunch thanks to the improved flexibility of its engine.

The 944S has now become something of a

Buying & Driving

For performance in the 944 era, the 250bhp Turbo S is in a class of its own, with a top speed of 162mph and 0-62mph acceleration in 5.7sec.

curiosity. Despite its 16-valve head, it lacked any real sparkle and did not achieve maturity until the 3-litre S2. As Porsche was working very hard from around 1988 onwards to tempt back customers, later cars had high levels of equipment to go with their stunning performance. The S2 then found new levels of success as a Cabriolet, and this version remains one of Porsche's most accomplished production sports cars, commanding a premium over an equivalent coupé.

The 944 Turbo can be grouped in that select band of production Porsches which are truly outstanding cars to drive. The first cars, available in 1985 and with 220bhp, were a revelation but, if buying one, watch out for some potentially expensive maintenance.

The best 944 Turbos are the 250bhp models, with the 1988 S – named the SE in the UK – a desirable limited edition. The specification for the latter, which included ABS and the Torsen limited slip differential, became the only 944 Turbo model in August 1988, but at lower cost. The range was then topped out by the stunning limited edition Turbo Cabriolet in 1991.

By any measure 944 Turbos are outstanding value today and, considering their price tags when they were new, later enthusiasts have the last laugh in terms of value for money.

The driving manners of the S2 are quite different to the Turbo. Its power is much smoother and the engine will pull easily from low revs. Whereas I might advise the Turbo to anyone wanting to take part in track days, mainly because of the way the power arrives in a big rush, the S2 is more driveable, although certainly a little slower in sporting use.

Always remember, though, that both the S2 and the Turbo are technically complex cars, where the service record usually holds the key.

Ten Tips when Buying a 944

1 Body The 944 series were fully galvanised from the first deliveries and came with the seven-year anti-corrosion guarantee, so it should be possible to find a car in excellent condition. Look out for overspray on the trim mouldings and number plate lamps. If you find the car has been resprayed, ask why. Are all the panels painted in the same shade and finish?

2 Engine oil The aluminium bores in 944 engines need consistent preventative maintenance in order to avoid excessive wear. Check for regular (6000-mile) oil and filter changes, using top quality oil. As the filter can be hard to remove

Porsche 924/944/968

because of the limited access to the front left side of the engine, changes may have been ignored in home servicing. Avoid cars with leaky engines.

3 Timing belts Check for evidence that the cam timing belts have been changed at maximum intervals of 36,000 miles, even though the factory now states that new belts are good for 45,000 miles. If a belt has been fitted recently, ensure its tension has been re-checked after 2000 miles if an automatic tensioner is not fitted.

Timing belt development can be tracked by a series of factory service bulletins, starting in December 1984. The main problem was wear and stretching, which required regular checks on belt tension, a very difficult mechanical job which required the use of a special tool. Owners of pre-1987 model year cars should check particularly for several improvements made by the factory. The first was the removal of the V-shaped support ribs on the rear of the plastic cover under the camshaft pulley, preventing belt contact and actioned on production cars made after April 1984, from engine numbers 43E11333 (manuals) or 43E21383 (automatics).

The tensioning pulley mounting was then strengthened to prevent it bending, resulting in a new oil pump body at the front of the engine on production engines after 43E14859/43E21784, from September 1984. From April 1985, at engine number 43F00071/43F20066, the belt itself was strengthened to improve the stretch margin, while from October that year new guidelines were laid down for setting and checking the belt tension with the special tool. This very complex manual adjustment then became history with the introduction of the automatic tensioner from the 928. All four-cylinder engines had this better automatic arrangement from the start of the 1987 model year.

The timing belts can be completely trouble-free as long as regular servicing is performed by a recognised expert specialist.

4 Oil/water mixing Is there evidence of oil mixing with water in the engine to form cream-coloured suds, visible on the oil dipstick, under the oil filler cap and looking down into the cam housing? This could mean a previously blown head gasket, which will be bad news if it has seriously overheated the all-aluminium engine, or possibly a broken oil/water cooler matrix, which is in turn a real pain to change.

5 Engine mounts The exhaust-side engine mount on the 944 was situated right next to the exhaust manifold, meaning early glycol-filled units would cook in certain extreme heat environments and then collapse. The symptoms were knocking or vibration when starting or shutting down, uneven idle running and difficult gear selection. The minimum height between the two mounting surfaces when the mount was installed was 60mm and if the dimension is less this means the mount is defective. Failure can also be seen from leaking fluid, or a wrinkled or cracked rubber surround on the mount.

After April 1985, from vehicle number 94EN460287, replacement with the stronger unit from the 944 Turbo was possible, the rubber surrounds of the new mounting painted grey for easy identification. Later items, stronger still, were painted orange (part number 944 37504202).

To reduce the effects of the heat further, 4mm spacers were fitted under the grey mounts to provide a degree of natural convection cooling over the body. The front exhaust-side louvre on the aluminium lower engine cover was also bent upwards to provide better air entry over the mounting, while the heat shield over the mounting in the engine compartment had a flap formed in its top surface to allow air to escape. Consult your Porsche specialist if you are unsure whether your early 944 has been given these modifications.

6 Clutch Replacement of the clutch involves dropping the transmission and torque tube back to remove the clutch disc, which can work out expensive, as it should involve fitting a new pressure plate and release bearing at the same time. When buying, look in the service records for evidence of clutch replacement after 50,000 miles, or as little as 30,000 miles on Turbos. Clutches on 944s often seem to have taken a beating from drivers unfamiliar with manual transmissions.

7 Front suspension Check the lower front suspension A-arms on pre-1985 models for signs of any damage, as this will cause 'tram-lining' or wander. Inspect alloy wheels carefully on both sides for evidence of 'kerbing', as replacements can be very expensive. Check tyres for even wear and steering alignment. Late '80s cars fitted with the optional ABS are very desirable.

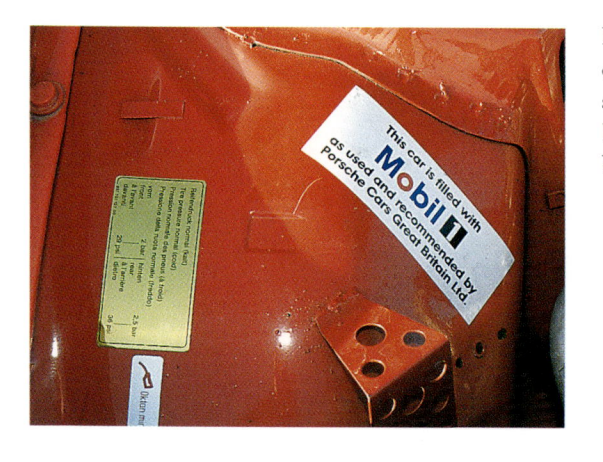

Regular renewal of high-quality oil, preferably synthetic, is vital for long life from the aluminium bores of 944 engines.

Buying & Driving

8 *Front wheel shimmy* When driving the car, check for evidence of front wheel shimmy or vibration. This is a typical 944 'rash' which is very difficult to eliminate and may well be caused by a number of individually small reasons which together add up to a problem. Usually it is the tyres, but it can be damaged rims, worn steering bushes, etc.

9 *History* Check for evidence of a consistent history. This does not have to be a full dealer history, but the service record or statutory annual inspections must give evidence of a consistent mileage build-up. Good repair shops always put a car's recorded mileage on their bills. The number of owners is not a guide, but the rule is still the fewer the better. And, of course, the higher the mileage, the lower the price should be.

10 *What about colour?* Unlike the 924, where the condition of the car is most important, there are still many 944s on the roads and when you come to resell you will find Guards Red the most saleable colour, closely followed by white and the metallics. Greens, browns and dark blues are the least wanted. Optional Fuchs wheels (the original five-spoke Porsche design), air conditioning and leather improve desirability.

Fourteen years after the 924's launch, Porsche finally introduced a Cabriolet. Early ones were rare 944S2 and 944 Turbo S versions, but this body style achieved deserved popularity during the 968 period.

Buying a 968

For Porsche the 968 may have been too little, too late, but the car still glittered with '90s technology and features. Even though it was over 80% new, many good features from the outgoing S2 were carried over – at the top of the list the inherent practicality the model had inherited from the earliest 924. The VarioCam engine had ample power and torque and, with its six-speed transmission, could match most high-performance supercars costing twice as much.

Choosing a 968 will be about what you want from your car. For luxury, go for the mainstream car in either coupé or Cabriolet form. For a track car for Europe, take the Club Sport.

All models have been hit hard by depreciation in their early years and represent good value, although not yet as good value as late-model 944s. But they have enough improvements to make the extra expenditure worthwhile, particularly on the desirable Cabriolet.

From about 1993 dealers were running out of ideas on how to sell the regular car, so look for those with lots of extras, including air conditioning, leather and metallic paint. The Sport for the British market, which lay half way between the regular car and the Club Sport, made the former

125

Porsche 924/944/968

obsolete overnight and certainly was not as impractical as the minimally-equipped Club Sport. Over time, there is little reason to doubt there will only be a small price difference to the regular car, making the latter a very good buy. Of course, you should remember the Club Sport was nearly 20% cheaper, which should now be reflected in the used price.

The Tiptronic automatic system was technically brilliant and later appeared on several other makes, including the better Audis, but as it did not sell too well on the 968 it is hard to find. If you like it, go for it, but be aware that performance will be degraded – although not as badly as on the original 924 automatic.

Ten Tips when Buying a 968

1 Shop around There are significant variations in price for equivalent models. Ignore people who tell you 968s do not depreciate as fast as 944s and look for 'loaded' models, especially those with options like metallic paint, air conditioning, cruise control, limited slip differential and Blaupunkt Symphony radio.

2 Service record Ensure the car has an immaculate one. This does not have to be with an official dealer, but must be with a reputable and known workshop. Look particularly for evidence of problems with the air conditioning.

Two-seater Club Sport trades practicality for purity in its sporting reflexes – and looks particularly wild in Speed Yellow.

Buying & Driving

Performance at a Glance

Year	Model	Weight (kg)	Power (bhp)	Top speed (mph)	0-62mph (sec)	Source
1976	924	1080	125	125	9.9	Factory
1976	924 (US)	1125	96[1]	118	11.8[2]	Various
1977.5	924 (US)	1125	110[1]	120	11.2[2]	Various
1977	924 auto	1235	125	121	11.4	Various
1978	924 auto (US)	1281	110[1]	115	12.4[2]	Road & Track
1979	924 Turbo	1180	170	140	7.8	Factory
1980	924 Turbo S (US)	1294	143[1]	134	9.3[2]	Road & Track
1980	924 Carrera GT	1180	210	150	6.9	Factory
1981	924 Turbo 2	1180	177	143	7.7	Factory
1981	924 Turbo 2 (US)	1200	154[1]	127	9.2[2]	Road & Track
1981	924 Carrera GTS	1121	245	155	6.2	Factory
1981	924 GTS Club Sport	1060	270	162	5.2	Factory
1981	924 Carrera GTR	945	375	181	4.7	Factory
1982	944	1180	163	131	8.4	Factory
1983	944 auto (US)	1315	143[1]	130	9.4[2]	Car & Driver
1985	944 Turbo	1350	220	152	6.3	Factory
1986	924S	1190	150	134	8.5	Factory
1987	924S auto	1240	150	134	9.8	Factory
1988	924S	1195	160	137	8.2	Factory
1987	944S	1280	190	142	7.9	Factory
1988	944	1260	160	136	8.4	Factory
1988	944 Turbo S	1350	250	162	5.7	Factory
1989	944 2.7	1290	165	137	8.2	Factory
1989	944S2	1310	211	149	6.9	Factory
1990	944S2 Cabrio	1390	211	149	7.1	Factory
1990	944 Turbo	1400	250	162	5.9	Factory
1991	944 Turbo Cabrio	1450	250	162	5.9	Factory
1992	968	1370	240	156	6.5	Factory
1992	968 Cabrio	1440	240	156	6.6	Factory
1992	968 Tiptronic	1430	240	154	7.9	Factory
1993	968 Club Sport	1320	240	160	6.3	Car
1993	968 Turbo S	1300	305	175	5.0	Factory

[1] Figure in SAE horsepower; all other figures are DIN.
[2] 0-60mph time.

3 Luxury If you are looking for this, aim for the later models which dealers loaded with options to move them out of their showrooms.

4 Crash damage As with any Porsche, check for this. Look for evidence of overspray where the doors shut, under the body and inside the front and rear compartments. A glass tailgate which is difficult to close may indicate distortion of the rear chassis.

5 Club Sport Only go for this model if you want a toy for weekends. It is useless for normal driving, but brilliant fun otherwise! Yellow or white are the most popular colours. If you want to avoid buying a converted racer, look behind the seats for evidence of a roll-over bar.

6 Cabriolet The Cabrio loses most of the legendary storage space of the coupé and is therefore not such a good choice if you have children. However, for open-air motoring it is obviously unbeatable. It was Porsche's best cabrio before the Boxster came along, although the keys for locking the front of the convertible top to the windscreen are a pain. Tall drivers will also complain about the difficulty of getting in and out. Cellphone installation would require some thought, as there are no hidden pockets in the rear of the car to hide the transceiver.

7 Colours Note the guidelines for best colours in the tips on the 944. For the regular coupé, the most saleable colours are the blue, black and grey metallics, although the darker metallics are good where being less conspicuous when parking is important. The Cabrio looks good in the lighter metallics like Horizon Blue and Polar Silver, as well as some of the wild colours.

8 Safety Airbags for both front occupants were available from the 1992 export launch, but not on right-hand drive cars until the start of the 1993 model year. ABS was standard on all models.

9 Wheels 968s have been fitted with several different types of tyres as original equipment, with the best fit for acceptable wear and grip probably the Michelin MXX3. Note that the optional 225/45ZR17 (front) and 255/40ZR17 (rear) sizes make the ride much firmer and are therefore not so suitable for general use.

10 Tiptronic For all its publicity, Tiptronic is still an automatic which adds some 20% to the 0-60mph acceleration time. But it is the best automatic around and the sequential push-pull gear change is fun. Go for it if you live in town or need two-pedal operation. Although they cost more than manuals when new, automatics have a history of depreciating more. If you haggle hard, you can get one for the same price, or less, than a manual version.

Tiptronic transmission was Porsche's technical master stroke of the '90s and sold particularly well on the 911. This popularity, especially in urban areas, will inevitably come to be reflected in values of 968 Tiptronics.

Porsche 924/944/968

ENGINE PERFORMANCE

	924	924 Turbo	924 Turbo	924CGT	924S	944	944	944 Turbo	944 Turbo	944S	944S2	968
Model year	1976	1978	1981	1980	1985	1982	1989	1986	1989	1987	1989	1992
Bore (mm)	86.5	86.5	86.5	86.5	100	100	104	100	100	100	104	104
Stroke (mm)	84.4	84.4	84.4	84.4	78.9	78.9	78.9	78.9	78.9	78.9	88	88
Capacity (cc)	1984	1984	1984	1984	2479	2479	2681	2479	2479	2479	2990	2990
Compression	9.3:1	7.5:1	8.5:1	8.0:1	9.7:1	10.6:1	10.9:1	8.0:1	8.0:1	10.9:1	10.9:1	11.0:1
Max power (bhp DIN)	125	170	177	210	150	163	165	220	250	190	211	240
Max torque (Nm)	165	244	250	275	195	205	225	330	350	230	280	305
Ignition[1]	Coil	TCIi	DME	DME	DME	DME	DME	DME	DME	DME	DME	DME
Injection	K-Jet	K-Jet	K-Jet	K-Jet	K-Jet	L-Jet	L-Jet	L-Jet	L-Jet	L-Jet	L-Jet	L-Jet

Notes
[1] 'Coil' refers to conventional ignition coil/contact breaker ignition; 'TCIi' refers to Bosch transistor breakerless ignition (for 924 US, for Series 1 924 Turbo and European 924s from 1980 model year); 'DME' refers to Digital Motor Electronics – called the Motronic system by its manufacturers, Bosch, from the later 944. All models are European specification. For reference, to convert torque from Nm to lb ft, divide by 1.356.

EVOLUTION OF WHEELS & TYRES

	924	924 Turbo	924 CGT	924S	944	944 Turbo	944 Turbo	944 S	944S2	968	968 Club Sport	968 Turbo S
Year	1976	1978	1980	1985	1982	1986	1989	1987	1989	1982	1993	1993
Unladen weight (kg)	1044	1204	1180	1164	1195	1258	1350	1280	1310	1370	1320	1300
Max power (bhp)	125	170	210	150	163	220	250	190	211	240	240	305
Wheel type	Steel	Cast alloy	Forged alloy	Cast alloy	Cast alloy	Cast alloy	Forged alloy	Cast alloy	Cast alloy	Cast alloy	Cast alloy	Cast alloy
Diameter (in)	14	15	15	15	15	16	16	15	16	16	17	18
Front rim width (in)	5.5	6	7	6	7	7	7	7	7	7	7.5	8
Rear rim width (in)	5.5	6	7	6	7	8	9	7	8	8	9	10
Front tyres	165HR	185/70VR	215/60VR	195/70VR	185/70VR[1]	205/55VR	225/50ZR	195/65VR	205/55ZR	205/55ZR	225/45ZR	235/40ZR
Rear tyres	165HR	185/70VR	225/50VR	195/70VR	185/70VR[1]	225/50VR	245/45ZR	195/65VR	225/50ZR	225/55ZR	255/40ZR	265/35ZR

Note
[1] Increased to 195/65VR for 1986 model year.

The right place to sign off: the UK-only 968 Sport, engineered like the inspired Club Sport but with added equipment and comfort, was the pinnacle of four-cylinder Porsche evolution.